# *Molecular Neurobiology for the Clinician*

**Review of Psychiatry Series**
John M. Oldham, M.D., M.S.
Michelle B. Riba, M.D., M.S.
*Series Editors*

# *Molecular Neurobiology for the Clinician*

EDITED BY

Dennis S. Charney, M.D.

No. 3

Washington, DC
London, England

**Note:** The authors have worked to ensure that all information in this book is accurate at the time of publication and consistent with general psychiatric and medical standards, and that information concerning drug dosages, schedules, and routes of administration is accurate at the time of publication and consistent with standards set by the U. S. Food and Drug Administration and the general medical community. As medical research and practice continue to advance, however, therapeutic standards may change. Moreover, specific situations may require a specific therapeutic response not included in this book. For these reasons and because human and mechanical errors sometimes occur, we recommend that readers follow the advice of physicians directly involved in their care or the care of a member of their family.



Manufactured in the United States of America on acid-free paper
07 06 05 04 03 5 4 3 2 1
First Edition

Typeset in Adobe's Palatino

American Psychiatric Publishing, Inc.
1000 Wilson Boulevard
Arlington, VA 22209-3901
www.appi.org

The correct citation for this book is

Charney DS (editor): *Molecular Neurobiology for the Clinician* (Review of Psychiatry Series, Volume 22, Number 3; Oldham JM and Riba MB, series editors). Washington, DC, American Psychiatric Publishing, 2003

**Library of Congress Cataloging-in-Publication Data**
Molecular neurobiology for the clinician / edited by Dennis S. Charney
p. ; cm.
Includes bibliographical references and index.
ISBN 1-58562-113-7 (alk. paper)
1. Neurobehavioral disorders—Molecular aspects. 2. Molecular neurobiology. 3. Biological psychiatry. I. Charney, Dennis S. II. Review of psychiatry series ; v. 22, 3.
[DNLM: 1. Biological Psychiatry. 2. Mental Disorders. 3. Neurobiology. WM 102 M718 2003]
RC455.4.B5M676 2003
616.89—dc21

2003043664

**British Library Cataloguing in Publication Data**
A CIP record is available from the British Library.

# Contents

# Contributors

**Dennis S. Charney, M.D.**
Chief, Mood and Anxiety Disorders Program; Chief, Experimental Therapeutics and Pathophysiology Branch, National Institute of Mental Health, Rockville, Maryland

**Edwin H. Cook Jr., M.D.**
Professor, Departments of Psychiatry, Pediatrics, and Human Genetics, University of Chicago, Chicago, Illinois

**Andrew R. Gilbert, M.D.**
Resident in Psychiatry, Department of Psychiatry, University of Pittsburgh School of Medicine, Pittsburgh, Pennsylvania

**Todd D. Gould, M.D.**
Postdoctoral Fellow, Laboratory of Molecular Pathophysiology, Mood and Anxiety Disorders Program, National Institute of Mental Health, Bethesda, Maryland

**Neil A. Gray, B.A.**
Predoctoral Fellow, Laboratory of Molecular Pathophysiology, Mood and Anxiety Disorders Program, National Institute of Mental Health, Bethesda, Maryland

**David A. Lewis, M.D.**
Professor, Departments of Psychiatry and Neuroscience, University of Pittsburgh School of Medicine, Pittsburgh, Pennsylvania

**Husseini K. Manji, M.D., F.R.C.P.C.**
Chief, Laboratory of Molecular Pathophysiology, Mood and Anxiety Disorders Program, National Institute of Mental Health, Bethesda, Maryland

**Francis J. McMahon, M.D.**
Chief, Genetic Basis of Mood and Anxiety Disorders, Mood and Anxiety Disorders Program, National Institute of Mental Health, Bethesda, Maryland

**Eric J. Nestler, M.D., Ph.D.**
Professor and Chairman, Department of Psychiatry and Center for Basic Neuroscience, The University of Texas Southwestern Medical Center, Dallas, Texas

**John M. Oldham, M.D., M.S.**
Professor and Chair, Department of Psychiatry and Behavioral Sciences, Medical University of South Carolina, Charleston, South Carolina

**Michelle B. Riba, M.D., M.S.**
Clinical Professor and Associate Chair for Education and Academic Affairs, Department of Psychiatry, University of Michigan Medical School, Ann Arbor, Michigan

**Jeremy Veenstra-VanderWeele, M.D.**
Department of Psychiatry, University of Chicago, Chicago, Illinois

**David W. Volk, Ph.D.**
Department of Neuroscience, University of Pittsburgh School of Medicine, Pittsburgh, Pennsylvania

# Introduction to the Review of Psychiatry Series

*John M. Oldham, M.D., M.S.*
*Michelle B. Riba, M.D., M.S., Series Editors*

**2003 Review of Psychiatry Series Titles**

- *Molecular Neurobiology for the Clinician*
  Edited by Dennis S. Charney, M.D.
- *Standardized Evaluation in Clinical Practice*
  Edited by Michael B. First, M.D.
- *Trauma and Disaster Responses and Management*
  Edited by Robert J. Ursano, M.D., and
  Ann E. Norwood, M.D.
- *Geriatric Psychiatry*
  Edited by Alan M. Mellow, M.D., Ph.D.

As our world becomes increasingly complex, we are learning more and living longer, yet we are presented with ever more complicated biological and psychosocial challenges. Packing into our heads all of the new things to know is a daunting task indeed. Keeping up with our children, all of whom learn to use computers and how to surf the Internet almost before they learn English, is even more challenging, but there is an excitement that accompanies these new languages that is sometimes almost breathtaking.

The explosion of knowledge in the field of molecular neurobiology charges ahead at breakneck speed, so that we have truly arrived at the technological doorway that is beginning to reveal the basic molecular and genetic fault lines of complex psychiatric diseases. In *Molecular Neurobiology for the Clinician*, edited by Dr. Charney, Dr. McMahon (Chapter 2) outlines a number of genetic discoveries that have the potential to affect our clinical practice in important ways, such as validating diagnostic systems and disease entities, improving treatment planning, and developing novel therapies and preventive interventions. Examples of these principles are illustrated in this book as they apply to addictive disorders (Chapter 4, by Dr. Nestler), schizophrenia (Chapter 3, by Dr. Gilbert and colleagues), psychiatric disorders of childhood and adolescence (Chapter 1, by Drs. Veenstra-VanderWeele and Cook), and mood and anxiety disorders (Chapter 5, by Dr. Gould and colleagues).

Increased precision and standardization characterizes not only the microworld of research but also the macroworld of clinical practice. Current recommendations regarding standardized assessment in clinical practice are reviewed in *Standardized Evaluation in Clinical Practice,* edited by Dr. First, recognizing that we must be prepared to reshape our diagnostic ideas based on new evidence from molecular genetics and neurobiology, as well as from the findings of clinical research itself. In Chapter 1, Dr. Basco outlines a number of problems inherent in routine clinical diagnostic practice, including inaccurate or incomplete diagnoses, omission of comorbidities, and various sources of bias, and an argument is made to train clinicians in the use of a standardized diagnostic method, such as the *Structured Clinical Interview for DSM* (SCID). Similar problems are reviewed by Dr. Lucas (Chapter 3) in work with child and adolescent patients, and a self-report diagnostic assessment technique, the Computerized Diagnostic Interview Schedule for Children (C-DISC), is described. The C-DISC is reported to have the advantages of enhancing patients' abilities to discuss their concerns and enhanced caretaker satisfaction with the intake interview.

Similarly, Dr. Zimmerman (Chapter 2) underscores the importance of developing a standardized clinical measure with good psychometric properties that could be incorporated into routine clinical practice, presenting data suggesting the value of one such system, the Rhode Island Methods to Improve Diagnostic Assessment and Services (MIDAS) project. Dr. Oquendo and colleagues (Chapter 4), in turn, review the critical issue of the use of standardized scales to enhance detection of suicidal behavior and risk of suicide in individual patients. The challenge to establish the cost-effectiveness of standardized assessment methodology in clinical practice is illustrated by the efforts in the U.S. Department of Veterans Affairs system, described by Dr. Van Stone and colleagues (Chapter 5), to train clinicians in the use of the Global Assessment of Functioning (GAF) scale and to incorporate it into the electronic medical record.

In *Trauma and Disaster Responses and Management,* edited by Drs. Ursano and Norwood, a compelling case is made by Dr. Bonne and colleagues (Chapter 1) of the fundamental interconnectedness between lifelong biological processes in humans and animals, and the environment. The relevance of studies using animal models to our understanding of posttraumatic stress disorder and other stress syndromes has become increasingly important, elucidating the functional neuroanatomy and neuroendocrinology of stress responses. This growing research database proves compelling when contemplating the human impact of major disasters such as the Oklahoma City bombing, described by Dr. North (Chapter 2), the effect of the World Trade Center tragedy and other disasters on developing children, described by Drs. Lubit and Eth (Chapter 3), and the potential and actual impact of bioterrorism on individuals and large populations, reviewed by Dr. Ursano and colleagues (Chapter 5). The need for early intervention, articulated by Dr. Watson and colleagues (Chapter 4), becomes increasingly clear as we learn more, which we must, about trauma and its effects.

As we learn more about many things, we make progress, but always with a cost. We are getting better at fighting illness and preserving health, hence we live longer. With longer life, however, comes new challenges, such as preserving the quality of life during these extended years. *Geriatric Psychiatry,* edited by Dr. Mellow, focuses on the growing field of geriatric psychiatry, from the points of view of depression (Chapter 1, by Dr. Mellow), dementia (Chapter 2, by Dr. Weiner), psychoses (Chapter 3, by Drs. Grossberg and Desai), late-life addictions (Chapter 4, by Drs. Blow and Oslin), and public policy (Chapter 5, by Dr. Colenda and colleagues). It is clear that we are making progress in diagnosis and treatment of all of these conditions that accompany our increased longevity; it is also clear that in the future we will increasingly emphasize prevention of illness and health-promoting habits and behaviors. Because understanding motivated behavior is a mainstay of what psychiatry is all about and we still have not unraveled all of the reasons why humans do things that are bad for them, business will be brisk.

Continuing our tradition of presenting a selection of topics in each year's Review of Psychiatry Series that includes new research findings and new developments in clinical care, we look forward to Volume 23 in the Review of Psychiatry Series, which will feature brain stimulation in psychiatric treatment (edited by Sarah H. Lisanby, M.D.), developmental psychobiology (edited by B.J. Casey, Ph.D.), medical laboratory and neuropsychiatric testing (edited by Stuart C. Yudofsky, M.D., and H. Florence Kim, M.D.), and cognitive-behavioral therapy (edited by Jesse H. Wright III, M.D., Ph.D.).

# Preface

*Dennis S. Charney, M.D.*

It is perhaps most appropriate to start this foreword with the question "Is molecular neurobiology even relevant to the clinician who is treating patients with mental illness?" In a recent task force report on the role of neuroscience in the future of psychiatric diagnosis, it was concluded that the next iterations of current DSM-IV must take into account advances in our understanding of the pathophysiology of major neuropsychiatric disorders (Charney et al. 2002). These advances will, in large part, be based on delineation of the molecular genetic contribution to the causes of psychiatric disorders; the molecular basis of neuronal network function, particularly in relation to abnormalities in cognitive and emotional regulation; and the identification of novel molecular targets for drug development. The chapters in this section provide a comprehensive view of the potential for molecular neurobiology to fundamentally change how we diagnose mental illness and treat patients.

The neurobiological basis of childhood- and adolescent-onset psychiatric disorders remains largely unknown. Drs. Veenstra-VanderWeele and Cook, in Chapter 1, nicely review the various research strategies, including genetic, postmortem, neuroimaging, and immunologic techniques, currently used. They also present the most important findings in patients with attention-deficit/hyperactivity disorder (ADHD), Tourette's syndrome, and obsessive-compulsive disorder (OCD).

The candidate gene approach focusing on the dopaminergic neural system has been most fruitful in the study of ADHD. Cook and colleagues (1995) originally reported transmission dysequilibrium at a variable number tandem repeat (VNTR) in the 3′ untranslated region of the dopamine transporter gene (DAT). Some, but not all, subsequent studies have replicated these findings, which is not surprising given the heterogeneity underlying the etiology of ADHD. Multiple research groups have also found a significant association at VNTR in exon 3 of the dopamine receptor $D_4$ gene *(DRD4)* (Faraone et al. 2001; LaHoste et al. 1996). It is not known if these genetic findings relate to clinical characteristics and treatment response.

Recent evidence suggests that the prevalence of autism has been increasing in the U.S. population (Yeargin-Allsopp et al. 2003). Autism appears to be a genetic disorder with a complex inheritance. As noted by Veenstra-VanderWeele and Cook, the pattern of relative risk in autism is consistent with multiplicative inheritance, with multiple genes interacting to form a heterogeneous phenotype. Although the pathophysiology and genetics of autism and related disorders remain elusive, the renewed commitment of the National Institute of Mental Health to autism research raises hopes for substantial progress in the near future.

The molecular genetic research strategies in Tourette's syndrome and childhood-onset OCD have involved both candidate gene and genomewide linkage and association studies. There is considerable familial overlap among Tourette's syndrome, tic disorders, and OCD, providing an additional challenge to the genetic studies. Unfortunately, neither the candidate nor the genomewide scans have led to replicated findings. The neuroimmunologic work by Swedo and colleagues (1997) suggesting a specific subtype of childhood OCD, designated pediatric autoimmune neuropsychiatric disorders associated with streptococcal infections (PANDAS), is a good example of the type of research needed to develop a pathophysiologically based diagnostic system.

Drs. Veenstra-VanderWeele and Cook make some important points at the conclusion of the chapter that bear repeating. They note that it will be critical to investigate the complex relationships among genetic variants, molecular neurobiology, and disease in the context of the genetic-environment interaction. The ultimate goal of molecular neurobiological research in child psychiatry is early identification of disease vulnerability and protective factors, development of preventative approaches, and the discovery of novel pharmacologic treatments.

In Chapter 2, Dr. McMahon reviews the current status—particularly the challenges—of molecular genetic research on complex adult-onset psychiatric disorders. He illustrates the very important point that bipolar disorder and schizophrenia likely involve at least two, but no more than four, genes. Furthermore, it appears likely that several genes collectively contribute to disease risk but that no one gene is sufficient. In addition, each susceptibility allele is very common. This profile is known in the field of human genetics as the *common disease–common variant hypothesis.* An important implication of this hypothesis is that even though disease susceptibility alleles are common and may individually increase risk for illness only slightly, they can have a large impact because of their frequency in the population.

Dr. McMahon concludes his chapter by prognosticating about the potential impact that genetic discoveries will have in psychiatry. He emphasizes how such discoveries will improve the validity and methodology associated with psychiatric diagnosis and lead to alterations in our diagnostic schema, promote the discovery of novel therapies, delineate with greater precision nongenetic causes of disease, and make prevention a real possibility. Of course, with the sequencing of the human genome and the advances in genetic techniques occurring almost daily, we all hope we get there sooner rather than later.

Dr. Lewis and his colleagues, in Chapter 3, focus on the relevance of recent molecular neurobiological findings to the etiology and treatment of schizophrenia. The etiology of schizophrenia clearly involves genetic factors, with the risk increasing with the percentage of shared genes. Most scholars agree that pathogenesis of schizophrenia involves the interplay of polygenetic influences and environmental risk factors, interacting with brain maturational processes (Lewis and Levitt 2002). However, it remains an important unknown as to whether schizophrenia is a single disease with a variety of phenotypic expressions or several different diseases with shared symptom patterns.

A couple of interesting candidate genes have emerged that may relate to specific elements of the schizophrenia process. Allelic variation in the gene for catechol-*O*-methyltransferase (COMT), a major enzyme involved in the metabolism of prefrontal cortex dopamine, may relate to impaired cognitive function in schizophrenia (Egan et al. 2001). The sensory gating abnormalities in schizophrenia may relate to a polymorphism on chromosome 15q14. The gene *(CHRNA7)* for the $\alpha_7$ subunit of the nicotinic acetycholine receptor (nACHR) is located near the chromosome 15 marker (Freedman et al. 1997; Weiland et al. 2000). The nACHR $\alpha_7$ subunit appears to be involved in the pathophysiology of gating deficits in schizophrenia and may account for the high rate of smoking by patients with schizophrenia. Smoking may be a form of self-medication with nicotine-stimulating nACHRs to reduce the subjective distress associated with sensory gating disturbances (Weiland et al. 2000).

The gene for RGS4, a member of the regulators of G-protein signaling proteins, is another schizophrenia candidate gene (Mirnics et al. 2001b). Postmortem microarray analysis revealed a decrease in cortical *RGS4* transcript in schizophrenia, and the *RGS4* gene is located on chromosome 1q21–22, a locus implicated in schizophrenia (Brzustowicz et al. 2000). A genetic association study that used RGS4 polymorphisms also supported altered *RGS4* expression as an inherited abnormality in schizophrenia (Chowdari et al. 2002). Finally, a significant association between *DTNBP1* (the gene for dystrobrevin-binding protein 1, or dysbindin) and schizophrenia has recently been reported (Straub et al. 2002). Dysbindin has been demonstrated to have a role in synaptic transmission, particularly involving γ-aminobutyric acid (GABA), supporting its candidacy as a susceptibility locus that influences the risk for schizophrenia.

What remains to be determined is how allelic variations in genes, such as those described above, relate to each other, to the abnormalities in brain circuitry identified, and to the spectrum of clinical phenotypes in schizophrenia. The synaptic-neurodevelopmental model of schizophrenia proposed by Mirnics and colleagues (2001a) has great appeal in this regard because it encompasses genetic, developmental, and biological elements of schizophrenia. Eventually, we hope that disease models such as this, which by analogy apply to other complex neuropsychiatric diseases, will lead to what we need most desperately: novel molecular targets for treatment intervention for schizophrenia.

Dr. Nestler, in Chapter 4, presents an extremely creative and potentially groundbreaking view of the molecular mechanisms and neural circuitry of reward and how they might relate to vulnerability to addictive behaviors. The foundation for Dr. Nestler's theory of the addicted state is that drug addiction is a form of drug-induced neural plasticity. Repeated exposure to a drug of abuse alters gene expression, which mediates altered function of individual neurons, which in turn fundamentally changes the neural networks of reward. As an example of this process, Dr. Nestler focuses on two transcription factors, CREB (cAMP response element binding protein) and ΔFosB.

The brain's reward and motivational neural system includes the nucleus accumbens (ventral striatum), ventral tegmental area, amygdala, hippocampus, lateral hypothalamus, and frontal cortex. Both CREB and ΔFosB are activated in the nucleus accumbens, perhaps the brain's primary reward region, but they mediate very different aspects of the addicted state. CREB mediates a form of tolerance as reflected by a reduction of sensitivity to repeated drug exposure. On the other hand, ΔFosB mediates sensitization to drug exposure and probably contributes to increased drug craving. Both CREB (e.g., dynorphin) and ΔFosB (e.g., GluR2) interact with many target genes to produce their effects.

Drug addiction is almost devoid of effective pharmacologic interventions that reduce behavioral sensitivity to drugs of abuse by reducing their rewarding properties. The work by Nestler and colleagues provides a theoretical framework from which novel putative "anti-addictive" drugs can be discovered.

In the final chapter, by Dr. Manji and colleagues, is a state-of-the-art review of the leading pathophysiological hypotheses of mood and anxiety disorders. These disorders are among the most common and disabling of all medical conditions. Yet, distressingly, the treatments used for these disorders use drugs with a mechanism of action and a clinical efficacy that are similar to those discovered decades before. This chapter moves beyond the classical monoamine theories of mood and anxiety disorders toward a molecular and cellular hypothesis that forms a basis for discovering the genetic basis of the different mood and anxiety disorders and for the strategic development of improved therapeutics. The authors discuss how abnormalities in complex signal pathways, and their effector elements, can account for the diverse symptomatology and pathophysiological findings in mood disorders.

The postulated protein kinase C, or PKC, signaling pathway mechanism of action of the mood stabilizers lithium and valproate suggests that PKC inhibitors, such as tamoxifen, may have antimanic properties. A large number of neuroimaging and postmortem investigations have revealed evidence of widespread morphological abnormalities in mood disorders. Dr. Manji and colleagues provide a framework from which to understand the etiology of these abnormalities, which point toward dysfunction in neurotrophic signaling molecules, including brain derived neurotrophic factor (BDNF) and bc1-2 (Manji and Chen 2002). Seminal studies by Ron Duman at Yale suggest that antidepressant drugs may have clinically relevant effects on neuronal atrophy and survival (D'sa and Duman 2002; Manji and Duman 2001).

The molecular and cellular hypothesis of mood disorders has led to many new targets for novel antidepressant and mood stabilizer drug development, including glutamate, glucocorticoid, corticotropin-releasing hormone, and cAMP-phosphodiesterase systems.

Dr. Manji and colleagues also review the advances made in our understanding of the neurobiology of anxiety disorders. Preclinical studies have defined with precision the neural circuitry and associated neurochemistry responsible for behaviors related to anxiety and fear. Elegant experiments by investigators such as Schafe et al. (2001), Myers and Davis (2002), and McGaugh and Izquierdo (2000) have resulted in new insights as to how the neural mechanisms related to fear conditioning and fear memory consolidation, retrieval, and reconsolidation are relevant to current conceptions of anxiety disorders such panic disorder and posttraumatic stress disorder. Clinical research investigations, especially those using neuroimaging and neuroendocrine techniques, have (more than ever before) defined the neural circuits and neurochemistry of human anxiety. This body of work has given impetus to the development and testing of new drug treatments for anxiety disorders, such as anxioselective benzodiazepine agonists, corticotropin-releasing hormone antagonists, and substance P antagonists. In addition, as illustrated in the chapter, many new candidate genes have emerged from these preclinical and clinical studies of anxiety and fear states.

Considered together, these chapters form an outstanding compendium providing the psychiatric clinician with a glimpse into the future of what increasingly will become a molecular-based discipline, similar to other medical specialties. I foresee a diagnostic system, within the next decade, that will become relevant to the pathogenesis of neuropsychiatric disorders. On the basis of the work conducted by the authors of these chapters and others, I foresee another "golden era" of pharmacotherapeutic drug development similar to that which occurred in the 1950s and 1960s. Finally, psychiatry is at the precipice of becoming a specialty in which the discovery of disease risk factors, based on the foundation built by molecular neurobiology and epidemiology research, will make prevention a major component of routine psychiatric practice.

## References

Brzustowicz LM, Hodgkinson KA, Chow EWC, et al: Location of a major susceptibility locus for familial schizophrenia on chromosome 1q21–q22. Science 288:678–682, 2000

Charney DS, Barlow DH, Botteron K, et al: Neuroscience research agenda to guide development of a pathophysiologically based classification system, in A Research Agenda for DSM-V. Edited by Kupfer DJ, First MB, Regier DA. Washington, DC, American Psychiatric Association, 2002, pp 31–83

Chowdari KV, Mirnics K, Semwal P, et al: Association and linkage analysis of RGS4 polymorphisms in schizophrenia. Hum Mol Genet 11: 1373–1380, 2002

Cook E, Stein M, Krasowski M, et al: Association of attention deficit disorder and the dopamine transporter gene. Am J Hum Genet 56:993–998, 1995

D'Sa C, Duman R: Antidepressants and neuroplasticity. Bipolar Disorder 4(3):183–194, 2002

Egan MF, Goldberg TE, Kolachana BS, et al: Effect of COMT Val 108/158 Met genotype on frontal lobe function and risk for schizophrenia. Proc Natl Acad Sci USA 98:6917–6922, 2001

Faraone SV, Doyle AE, Mick E, et al: Meta-analysis of the association between the 7-repeat allele of the dopamine $D_4$ gene and attention deficit hyperactivity disorder. Am J Psychiatry 158:1052–1057, 2001

Freedman R, Coon H, Myles-Worsley M, et al: Linkage of a neurophysiological deficit in schizophrenia to a chromosome 15 locus. Proc Natl Acad Sci USA 94:587–592, 1997

LaHoste G, Swanson J, Wigal S, et al: Dopamine D4 receptor gene polymorphism is associated with attention deficit hyperactivity disorder. Mol Psychiatry 1:128–131, 1996

Lewis DA, Levitt P: Schizophrenia as a disorder of neurodevelopment. Annu Rev Neurosci 25:409–432, 2002

Manji HK, Duman RS: Impairments of neuroplasticity and cellular resilience in severe mood disorder: implications for the development of novel therapeutics. Psychopharmacol Bull 35(2):5–49, 2001

Manji HK, Chen G: PKC, MAP kinases and the bcl-2 family of proteins as long-term targets for mood stabilizers. Mol Psychiatry 7 (suppl 1): S46–S56, 2002

McGaugh JL, Izquierdo I: The contribution of pharmacology to research on the mechanisms of memory formation. Trends Pharmacol Sci 21(6): 208–210, 2000

Mirnics K, Middleton FA, Lewis DA, et al: Analysis of complex brain disorders with gene expression microarrays: schizophrenia as a disease of the synapse. Trends Neurosci 24:479–486, 2001a

Mirnics K, Middleton FA, Stanwood GD, et al: Disease-specific changes in regulator of G-protein signaling 4 (RGS4) expression in schizophrenia. Mol Psychiatry 6:293–301, 2001b

Myers KM, Davis M: Behavioral and neural analysis of extinction. Neuron 364(4):567–584, 2002

Schafe GE, Nader K, Blair HT, et al: Memory consolidation of Pavlovian fear conditioning: a cellular and molecular perspective. Trends Neurosci 24:540–546, 2001

Straub RE, Jiang Y, MacLean CJ, et al: Genetic variation in the 6p22.3 gene DTNBP1, the human ortholog of the mouse dysbindin gene, is associated with schizophrenia. Am J Hum Genet 71(2):337–348, 2002

Swedo SE, Leonard HL, Mittleman BB, et al: Identification of children with pediatric autoimmune neuropsychiatric disorders associated with streptococcal infections by a marker associated with rheumatic fever. Am J Psychiatry 154:110–112, 1997

Weiland S, Bertrand D, Leonard S: Neuronal nicotinic acetylcholine receptors: from the gene to the disease. Behav Brain Res 113:43–56, 2000

Yeargin-Allsopp M, Rice C, Karapurkar T, et al: Prevalence of autism in a US metropolitan area. JAMA 289:49–55, 2003

Chapter 1

# Molecular Neurobiology of Childhood- and Adolescent-Onset Psychiatric Disorders

*Jeremy Veenstra-VanderWeele, M.D.*
*Edwin H. Cook Jr., M.D.*

The advent of the Human Genome Project holds much promise for the field of psychiatry and for child psychiatry in particular. Rapid advances in molecular genetic technology have allowed researchers to hope for definitive answers emerging out of large collaborative linkage studies. Mapping of chromosomal abnormalities and linkage studies have already identified genes in a number of rare simple genetic disorders with child psychiatric phenotypes. Large consortia are moving rapidly in the search for linkage in common childhood-onset psychiatric disorders. While these consortia were forming, association studies using a candidate gene approach had already implicated a number of genes.

Genetic findings are emerging rapidly, but molecular neurobiology extends beyond genetics. Postmortem studies are now being used to evaluate differential expression of protein or mRNA. These initial studies may eventually lead to protein or mRNA array techniques that may identify protein systems and signaling cascades disrupted in childhood neuropsychiatric diseases. As neuroimaging technology progresses, brain scans using positron emission tomography (PET), single photon emission computed tomography (SPECT), and magnetic resonance spectroscopy (MRS) will assist in the study of neurochemical and neuroreceptor abnormalities in vivo. Investigators are exploring the molecular basis of possible autoimmunity in Tourette syndrome and pediatric-onset obsessive-compulsive disorder (OCD). Animal models based on simple genetic diseases or behavioral phenotypes allow exploration of molecular changes at the synaptic level. In the case of the pervasive developmental disorder Rett syndrome, the animal model has been built on a specific genetic abnormality.

Though promising, the emergence of novel molecular technology should not distract us completely from some of the molecular puzzles described in previous decades. Hyperserotonemia in autism remains an incomplete story. Endocrine abnormalities in early-onset depression continue to be tantalizing clues. These early findings may be explained once mRNA and protein approaches have been fully explored, but more direct investigation may yield quicker resolution of the mystery—and may produce important information about the disorders themselves.

In this chapter we discuss the breadth of molecular approaches in child neuropsychiatric disease. Unfortunately, child psychiatric research has lagged behind adult research in funding and application of molecular techniques, so less depth of research is available in each area.

The chapter is organized by illness, with rare chromosomal and simple genetic diseases presented first. The more common illnesses, presented second, are thought to have complex etiology, and thus multiple molecular techniques are typically being brought to bear on each one. Rather than beginning the chapter with an independent explication of each concept and technique, we have opted to discuss these approaches in the context of disease. This allows discussion of an approach's strengths and weaknesses using real examples. For example, the discussion of Prader-Willi and Angelman syndromes explicates the concept of genetic imprinting. The first section on a complex genetic disease, attention-deficit/hyperactivity disorder (ADHD), includes a discussion of statistical approaches in molecular genetics. Likewise, the section on Tourette syndrome and childhood-onset OCD describes the immunological techniques that have been used in studying childhood psychiatric disorders.

## Chromosomal and Relatively Simple Genetic Disorders

Unlike most people with psychiatric disorders with complex genetic and environmental etiologies, persons with chromosomal or relatively simple genetic disorders can point to a single mutation or chromosomal abnormality as contributing most of the risk for their illness. Within each relatively simple genetic disorder, such as early-onset familial breast cancer caused by mutations in *BRCA1,* penetrance and expression may vary. Variable expression can be seen in women with mutations in *BRCA1,* since some may develop ovarian cancer or other types of cancer instead of breast cancer (Friedman et al. 1995). Variable penetrance is also evident, as there are some women with *BRCA1* mutations who never develop breast cancer. Penetrance is a particularly relevant concept in child psychiatry, where penetrance rates may depend on age.

A few disorders with onset in childhood, such as Rett syndrome, fragile X syndrome, and maternally inherited chromosome 15q11–q13 duplication and triplication, are relatively simple genetic disorders. More single-gene or chromosomal disorders will likely be identified from within heterogeneous syndromes such as autism; however, simple genetic disorders will not account for the majority of children and adolescents with genetic susceptibility to neuropsychiatric illness.

Chromosomal abnormalities may provide important clues to genes and protein systems relevant in child psychiatry. A few of the more common chromosomal disorders are discussed below. Rare chromosomal anomalies associated with specific psychiatric disorders, such as chromosome 15q duplications and inversions in autism, are discussed in the complex genetic disorders section.

Simple genetic disorders are rare in child psychiatry, as in other medical disciplines (except clinical genetics), and they often include both psychiatric problems and other clinical features. A number of relatively simple genetic syndromes primarily affect other systems but also have prominent child psychiatric phenotypes. Smith-Lemli-Opitz syndrome (SLOS), which results from a defect in cholesterol biosynthesis, typically presents with characteristic facies, microcephaly, hypotonia, syndactyly, and hypogenitalism; however, SLOS also prominently includes self-injurious behavior and symptoms of autism in a majority of children (Tierney et al. 2001). Lesch-Nyhan syndrome results from a purine salvage pathway defect and typically presents with choreoathetosis, dystonia, compulsive self-injury, and mental retardation (Nyhan 1997). The manifestations of tuberous sclerosis (TSC) include growth of multiple hamartomas, mental retardation, seizures, and often symptoms of autism. TSC results from defects in one of two tumor suppressor genes, *TSC1* and *TSC2,* which encode hamartin and tuberin, respectively, and the resulting tubers may account for the symptoms of autism in these patients (O'Callaghan and Osborne 2000). Unlike these and other simple genetic syndromes that include child psychiatric manifestations as a smaller part of their overall phenotype, Rett syndrome and fragile X syndrome manifest predominantly as neuropsychiatric disorders.

### *Rett Syndrome*

Rett syndrome (RTT) is an X-linked disorder with an incidence of approximately 1 in 10,000–15,000 girls, with 99.5% of cases being sporadic. Girls with RTT develop normally until approximately 6–18 months of age, when they lose speech and purposeful hand movements and then commonly progress to develop microcephaly, seizures, ataxia, stereotypic hand movements, abnormal breathing patterns, and autistic behavior. Molecular markers of cholinergic neurons, including the enzyme choline acetyltransferase, are lower in RTT (Wenk and Mobley 1996).

Mutations in *MECP2,* the gene encoding methyl-CpG binding protein 2 (MeCP2), were recently identified as the cause of RTT (Amir et al. 1999). The MeCP2 protein normally binds methylated CpG dinucleotides, thereby forming part of the Sin3A/HDAC complex that deacetylates histones to repress gene expression. Therefore, RTT must somehow be a result of inappropriate gene overexpression at some time and location during development. Heterozygous female *mecp2* knockout mice develop symptoms at about 9 months of age, after reaching maturity, which raises the possibility that the genetic defect could affect brain stability rather than development (R. Z. Chen et al. 2001; Guy et al. 2001) and is consistent with the finding that MeCP2 protein is expressed after neurons reach a certain level of maturity (Shahbazian et al. 2002a). Male mice heterozygous for an *mecp2* truncation mutation have symptoms developing at 6 weeks and eventually manifest many of the core features of Rett syndrome, including even stereotyped forelimb movements, less exploration of novel environments (open-field test), and social dysfunction (Shahbazian et al. 2002b). In a possible model of the human stigmatization of those with developmental disorders and mental illness, wild-type mice spend less time interacting with the mutant mice when placed in the same cage.

Once *MECP2* mutations were identified in Rett syndrome, the involvement of the gene was established in other conditions as well. Males who receive a mutated form of *MECP2* in families with classic RTT may survive into the neonatal period but appear to have severe encephalopathy that frequently leads to death (Wan et al. 1999). In contrast to typical RTT, some missense mutations appear to characteristically lead to either no phenotype or mild mental retardation with some characteristics of RTT in females and to moderate to severe mental retardation in males (Orrico et al. 2000).

## *Fragile X Syndrome (FRAXA)*

Fragile X syndrome (FRAXA) due to defects in *FMR1* function is the prototypical genetic syndrome caused by an expanding trinucleotide repeat sequence. The typical clinical picture in FRAXA includes mental retardation, macroorchidism, large ears, prominent jaw, and high-pitched, jocular speech. Increased rates of autism spectrum disorders and ADHD are also observed in FRAXA. While interest is appropriately focused on the relatively high rate of autistic features in fragile X syndrome patients, the rate of FRAXA in the population of referred psychiatric patients with autism is very low. In the vast majority of cases, the responsible gene variant is an expanding CGG triplet repeat variant that interferes with transcription of *FMR1* at chromosome Xq27. Expanding triplet repeats at this gene can increase in number when transmitted from parent to child, eventually exceeding a threshold of 200 repeats, above which *FMR1* is hypermethylated and the protein is no longer expressed. Rare cases of missense mutations have been reported. The exact role of the FMR1 protein (FMRP) in neurons is not clear, but FMRP interacts with mRNAs and ribosomes, suggesting a role in protein synthesis. FMRP is heavily synthesized in dendritic spines in response to synaptic activity, and abnormal dendritic spine size and shape have been noted in FRAXA patients and *fmr1* knockout mice (Greenough et al. 2001). Furthermore, *fmr1* knockout mice have learning deficits and show hyperreactivity to sound (Chen and Toth 2001). As an X-linked disorder, FRAXA typically affects boys, but it may also affect girls to a milder degree, and the level of FMRP expression in girls may correlate with the degree of behavior problems (Hessl et al. 2001).

## *22q11 Deletion Syndrome*

Deletion of chromosome 22q11 is associated with multiple clinical manifestations that have been variously labeled velocardiofacial syndrome (VCFS), CATCH22, and 22q11 deletion syndrome. The associated anomalies include typical facies; learning disabilities; mental retardation; cardiac defects; and multiple midline defects, including cleft palate and brain malformations. Recent work has focused on both the occurrence of psychiatric disorders in children and adults with the 22q11 deletion syndrome and the prevalence of undiagnosed 22q11 deletion syndrome in patients with previously diagnosed psychiatric disorders. In a sample of children and adolescents with 22q11 deletion syndrome, elevated rates of major depression, dysthymia, and the trait of hyperactivity/inattention were seen in comparison to their siblings (Arnold et al. 2001). The incidence of schizophrenia is as high as 24% in adult patients with 22q11 deletion syndrome (Murphy et al. 1999), and bipolar disorder also occurs at elevated rates (Carlson et al. 1997). Investigators have also found that 22q11 deletions are detected at an elevated rate in both children and adults with schizophrenia (Karayiorgou et al. 1995; Usiskin et al. 1999).

## *Prader-Willi and Angelman Syndromes*

Prader-Willi and Angelman syndromes both reflect genetic imprinting of a region of chromosome 15. Prader-Willi syndrome typically results from either deletion of the paternal chromosome 15q11–q13 region or complete lack of the paternal 15th chromosome with two otherwise normal copies of the maternal chromosome (maternal uniparental disomy). The clinical features of Prader-Willi syndrome include hypotonia, mild mental retardation, short stature, hypogonadism, and compulsive eating behavior with obesity. The disorder is thought to result from the imprinting of genes that are expressed only from the paternal chromosome and are therefore not expressed in Prader-Willi syndrome. Although previous dogma has emphasized the protein-coding function of mRNA, the leading candidate gene in Prader-Willi syndrome is an antisense transcript that encodes multiple snoRNAs, some of which are complementary to the serotonin 5-$HT_{2C}$ receptor gene and may mediate RNA editing (Cavaille et al. 2000). In contrast, Angelman syndrome has four mechanisms: mutation in *UBE3A* (Matsuura et al. 1997), a gene important in protein degradation; deletion of the maternal chromosome 15q11–q13 region; complete lack of the maternal 15th chromosome (paternal uniparental disomy); or imprinting mutations (in which the switch in methylation pattern of a mother's chromosome inherited from her father does not take place on transmission to her child).

Angelman syndrome is characterized by moderate to profound mental retardation, ataxia, hypotonia, characteristic facies, epilepsy, absence of speech, and predominant smiling and laughter. In contrast to Prader-Willi syndrome, which appears to be dependent on disruption of multiple genes, Angelman syndrome is thought to result largely from lack of expression of the maternally expressed *UBE3A* gene in the brain; however, patients with deletions or uniparental disomy have more severe phenotypes, suggesting that deletion of other genes, such as the γ-aminobutyric $acid_A$ ($GABA_A$) receptor subunit genes (*GABRB3, GABRA5, GABRG3*), may contribute to the more severe phenotype in deletion cases (Moncla et al. 1999).

### *Turner Syndrome*

Turner syndrome affects females with only one copy of the X chromosome (45,X). The typical clinical features include short stature; webbing of the posterior neck; increased carrying angle of the arms; sternal deformities, including pectus excavatum; and streak ovaries. Some individuals with Turner syndrome also show social and cognitive impairments and have an elevated risk of autism and ADHD. Analysis of parental origin of the X chromosome revealed that girls with an X chromosome from their mother (45, $X^{m}$) (i.e., in whom the X chromosome from the father was absent) had significantly worse social cognition, behavioral inhibition, and verbal and visuospatial memory than those who received their X chromosome from their father (45, $X^{p}$) (Skuse et al. 1997). The parent-of-origin effect suggests an imprinted locus (a gene on the X chromosome expressed only when inherited from the father) that could have an effect on differences between male and female psychosocial development, since males do not receive an X chromosome from their father. Such an effect could be partially responsible for male predominance of almost all psychiatric disorders of prepubertal onset.

## Complex Disorders

Most child and adolescent psychiatric disorders have complex etiology. For each disorder, etiology can be divided into genetic and environmental effects that have been classically debated as "nature versus nurture." Environmental effects do not, however, simply represent parenting and other components of the psychosocial equation. Environmental effects might include placental factors, infection, head trauma, or toxins. Gene-environment interactions may also be important. One classic example of this would be rheumatic fever, in which a familial risk interacts with pharyngeal infection by group A streptococcus.

Environmental effects are characteristically difficult to study. Controlled, experimental studies of potential environmental causes are ethically and logistically impossible in humans, so investigators are forced to adopt clever, but limited, epidemiological approaches to identify potential risk factors. Rarely, clinical or epidemiological observations can be adapted into animal models that may allow experimental design.

Other molecular approaches do not clearly differentiate environmental or genetic effects but seek to establish traits that characterize the disorder itself in the hope that they may clarify etiology. Examples of this approach include studies of hyperserotonemia in autism, B cell markers in Tourette syndrome and pediatric-onset OCD, and endocrine markers in pediatric-onset major depressive disorder.

Genetic approaches in complex disorders seek genetic variants that converge to increase susceptibility to a given disorder. There are likely rare situations in which a patient has a simple genetic disorder that cannot yet be distinguished from the overall syndrome; however, these "subsyndromes" will only represent a small fraction of patients with the disorder. Since for most patients, multiple genetic variants must occur together to produce disease, each variant is likely to be frequent in the overall population. Instead of describing these common variants as "mutations" or "defects," they are more appropriately called "variants," to reflect the notion that they do not cause disease except in interaction with other genetic or environmental factors. No complex genetic disorders have been parsed to identify all susceptibility genes, but some potential patterns of gene interactions have been proposed, including additive and multiplicative effects. Patterns of disease within relatives of probands are used to model types of gene-gene interactions.

While the simple genetic disorders clearly represent heterogeneity within a given child psychiatric syndrome, two types of genetic heterogeneity are important for the larger group of patients with a complex genetic disorder. *Locus heterogeneity* means that defects in different loci, or genes, may cause the same phenotype. The concept arose in relatively simple genetics, in which, for example, defects in one of a few different genes can cause hereditary nonpolyposis colorectal cancer (Nystrom-Lahti et al. 1994). This idea becomes much more complicated in complex genetic disorders, in which different (likely overlapping) groups of genetic variants may cause susceptibility to disease. On the other hand, *allelic heterogeneity* means that different defects in the same gene may lead to different patterns of genetic disease. This concept also arose in simple genetic disorders, in which different mutations in the same gene can lead to different phenotype patterns. For example, many different mutations in a single gene *(CFTR)* have been described in cystic fibrosis, some of which are associated with particularly mild forms of the disorder or may lead to more pancreatic than lung findings (The Cystic Fibrosis Genotype-Phenotype Consortium 1993). As rare and common variants in relevant genes (e.g., the serotonin transporter gene *[SLC6A4]*) are better understood, allelic heterogeneity may also become relevant in complex genetic disorders in psychiatry.

The various molecular approaches used in complex disorders, with their strengths and weaknesses, will be described with the first disorder in which they have been incorporated. Certain approaches have been used in only a limited number of the disorders, and these will typically be described only in cases in which they are clearly relevant. For example, potential immunologic approaches will not be discussed in ADHD, since no evidence suggests that they are important. A number of psychiatric disorders have variable age at onset, and these will be discussed below only when the pediatric-onset disorder represents a substantial fraction of the total patients with the disorder.

## *Attention-Deficit/Hyperactivity Disorder*

Attention-deficit/hyperactivity disorder is the most common childhood psychiatric disorder, with prevalence rates of 4%–11% in school-age children and much higher prevalence in males than in females. Molecular studies in ADHD have consisted of neurochemical and genetic approaches. Both approaches have thus far centered primarily on the dopaminergic system.

Despite significant efforts, few studies have identified abnormalities in dopamine itself, its principal metabolite homovanillic acid (HVA), or other catecholamines and their metabolites either centrally, in the cerebrospinal fluid (CSF), or peripherally, in the plasma or urine. Studies of urinary or plasma catecholamines in ADHD have not revealed a consistent relationship with the disorder or symptomatology (Pliszka et al. 1996). Some studies, but not all, show a correspondence between ADHD symptoms and CSF HVA levels (Castellanos et al. 1994, 1996; Zametkin and Rapoport 1987).

In vivo studies of receptor and transporter binding have recently become possible with SPECT and PET. In some, but not all, studies, these methods identified increased striatal dopamine transporter availability in ADHD (Dougherty et al. 1999; Dresel et al. 2000; Krause et al. 2000; van Dyck et al. 2002). Conversely, methylphenidate decreases dopamine transporter availability (Dresel et al. 2000; Krause et al. 2000; Volkow et al. 2002) and increases extracellular dopamine as assessed by a decrease in dopamine $D_2$ receptor availability (Volkow et al. 2001). Single studies have shown increased conversion and storage of labeled dihydroxyphenylalanine (DOPA) in the prefrontal cortex in adults (Ernst et al. 1998) and in the midbrain in children (Ernst et al. 1999).

Animal studies have long been used to test and generate hypotheses about ADHD. Initial lesion studies destroyed dopamine neurons in developing rats with the toxic 6-hydroxydopamine compound and identified accompanying increases in activity levels (Shaywitz et al. 1976). Interestingly, antagonists to dopamine receptor $D_4$ attenuate this hyperactivity (Zhang et al. 2001). Later molecular efforts have focused on mice that are hyperactive either by breeding or by genetic engineering. Mice heterozygous for the coloboma mutation were so named for ocular dysmorphology, but they also demonstrate spontaneous hyperactivity responsive to D-amphetamine but not methylphenidate (Wilson 2000). The coloboma mutation is a chromosomal deletion encompassing a number of genes, including *snap*, which encodes SNAP-25, a synaptic vesicle docking protein that appears to be particularly important in presynaptic regulation of dopamine and norepinephrine release (Jones et al. 2001). When *snap* is transgenically returned to this mouse lineage, the mice return to normal levels of activity (Hess et al. 1996).

An inbred rat model is the most extensively studied animal model for ADHD. Spontaneously hypertensive rats (SHRs), bred to manifest hypertension, also demonstrate marked hyperactivity as juveniles that decreases with administration of methylphenidate or D-amphetamine (Sagvolden et al. 1993). Molecular studies in these rats have revealed complex alterations of the dopaminergic system (Papa et al. 2000; Russell et al. 1995) and of nicotinic acetylcholine receptors (Terry et al. 2000). Interbreeding of the SHR and its originating Wistar-Kyoto (WKY) strain allowed localization of a quantitative trait locus (QTL) for hyperactivity on rat chromosome 8 (Moisan et al. 1996), but the gene responsible has not yet been identified.

Genetic knockout and knockdown studies are now being used to test hypotheses about particular genes in ADHD. As a contrast to the initial lesion findings, hyperdopaminergic mice with no dopamine transporter *(DAT)* gene expression or decreased *DAT* expression manifest dramatic hyperactivity that diminishes with administration of D-amphetamine or methylphenidate (Giros et al. 1996; Zhuang et al. 2001). Since methylphenidate is typically thought to act primarily by blocking the dopamine transporter, this surprising finding suggests that it also acts through the serotonergic system (Gainetdinov et al. 1999).

Human genetic studies in ADHD are complicated by sibling contrast effects in parent ratings, in which parents tend to rate siblings in contrast to one another, rather than in contrast to the general population (Eaves et al. 2000). Twin studies have found a monozygotic concordance rate of 51%–58% and a dizygotic concordance rate of 31%–33% (Goodman and Stevenson 1989; Sherman et al. 1997). The recurrence rate in first-degree relatives is approximately 25%, with higher rates in males (Biederman et al. 1992). Higher recurrence rates are identified when subtypes of ADHD are considered, particularly ADHD with comorbid conduct disorder or ADHD that persists into adolescence (Faraone et al. 2000). Using subtypes with a higher genetic loading may increase power to detect association or linkage (J. Holmes et al. 2002).

The genetic studies in ADHD provide a representative picture of genetic approaches in child and adolescent psychiatry. The two statistical approaches used in child and adolescent psychiatric disorders are linkage and association studies. With a few notable exceptions, tests of genetic association have been used to study candidate genes in disease. Tests of association are considerably more powerful than tests of linkage at a given locus, allowing genes of weaker effect to be detected. These tests compare the observed alleles in individuals with a disease with those expected by chance. Association tests may be performed by comparing allele ratios in patients with those in control subjects, but this case-control method is vulnerable to the potential bias of population stratification. This bias occurs when the case and control populations differ subtly in their ethnic (genetic) origins such that differences between them do not necessarily reflect allelic association with disease. Family-based association testing avoids this bias and has become the gold standard. More recently, methods that properly control for potential population stratification (genome control and associate) have been developed for case-control studies (Bacanu et al. 2000; Pritchard and Donnelly 2001).

The candidate gene approach has proven more fruitful in child than in adult psychiatry, and particularly so in ADHD. This approach is dependent on some understanding of the biology of a disorder. In ADHD, this understanding is based on response to stimulants, and genes within the dopaminergic system have been heavily researched. Cook and colleagues (1995) originally reported transmission disequilibrium at a variable number tandem repeat (VNTR) in the 3′ untranslated region (UTR) of the dopamine transporter gene *(DAT)*. Some, but not all, subsequent studies have also found that the 10-copy version of this polymorphism is preferentially transmitted from heterozygous parents to their affected children. Recent evidence demonstrates that the VNTR can act as a transcriptional enhancer (Michelhaugh et al. 2001). However, Barr and colleagues (2001) found significant transmission disequilibrium only when two additional polymorphisms within *DAT* were included in an extended haplotype, suggesting that an unidentified susceptibility variant may also be involved, whether within the VNTR or elsewhere in the gene.

Multiple studies in ADHD have also found significant association at a VNTR polymorphism in exon 3 of the dopamine receptor $D_4$ gene *(DRD4)* that has also shown significant association with the trait of novelty seeking in some samples (Ebstein et al. 1996). The 7-copy repeat, which leads to a less responsive version of the receptor (Asghari et al. 1995), was first found, in a study using a case-controlled design, to occur at a higher frequency in ADHD (LaHoste et al. 1996), and follow-up family-based association studies have now replicated the association (Faraone et al. 2001). Despite the functional nature of the exon 3 polymorphism, other polymorphisms could also be involved, and evidence for transmission disequilibrium has also been found at a promoter repeat element (McCracken et al. 2000).

Other associations have also been reported in ADHD. A few groups have found transmission disequilibrium at a polymorphism near the dopamine receptor $D_5$ gene *(DRD5)* (Daly et al. 1999). Following up on the coloboma mutant mouse findings, two studies have found transmission disequilibrium at separate markers in the SNAP-25 gene in ADHD (Barr et al. 2000; Mill et al. 2002). Family-based associations have also been reported for the dopamine β-hydroxylase gene *(DBH)*, the serotonin receptor 5-$HT_{2A}$ gene *(HTR2A)*, the monoamine oxidase A gene *(MAOA)*, and the DXS7 locus in linkage disequilibrium with the monoamine oxidase genes, but these findings have yet to be replicated.

While the candidate gene approach has proven fruitful in ADHD, it has a large drawback. We have only a very limited understanding of the pathophysiology of any psychiatric disorder, so the number of potential candidate genes in a given disorder is large, given the large number of genes expressed during brain development. In contrast to candidate gene association studies, genomewide approaches using linkage mapping do not require a priori knowledge of the pathophysiology of a disorder. Linkage statistics compute sharing of a chromosomal region among members of a family affected with a disorder. On the basis of experience in adult psychiatric disease, investigators currently favor nonparametric approaches that make no assumptions about disease transmission. These approaches typically use affected sibling pairs to identify increased sharing of alleles at particular points in the genome, but similar approaches may be applied to larger families when available.

The first genomewide study in ADHD did not initially yield any regions with evidence for linkage that achieved genomewide significance (Fisher et al. 2002b). Two qualitative approaches to sibling pairing were used: "broad ADHD," including probable and definite diagnoses; and "narrow ADHD," including only definite diagnoses. One quantitative measure was applied on the basis of the number of positive symptoms in the both the DSM-IV inattention and hyperactivity criteria, and this measure corresponded primarily to score in the hyperactivity dimension. Subjects were not separated into groups with hyperactive or inattentive subtypes due to sample size limitations (126 pairs). One region on chromosome 12p13 showed suggestive evidence for linkage in the quantitative analysis, but major loci could be excluded from 96% of the genome. A follow-up study genotyped a region on chromosome 16p in a larger set of sibling pairs and found significant evidence for linkage at chromosome 16p13 (Smalley et al. 2002), a region also implicated in some studies in autism. Since hyperactivity and inattention are common in autism, it is tempting to believe that a common genetic susceptibility variant may contribute to either disorder, with diagnostic specificity related to the number of other genetic and environmental risk and protective factors. Larger sample sizes will enable subtyping to reduce heterogeneity and may allow other linkage peaks to be identified. Quantitative trait approaches may be quite helpful but will likely need to use more extensive behavior rating scales than simply the DSM-IV criteria. More sophisticated quantitative trait analyses are already being explored in association studies (Curran et al. 2001; Waldman et al. 1998).

Thus far, molecular studies in ADHD have implicated the dopaminergic system, but the exact perturbations of this system remain frustratingly unclear. Mice with lesions of their dopaminergic neurons manifest hyperactivity, but so do mice with reduced uptake of dopamine. A number of genes within the dopaminergic system have been implicated in more than one study, but their specific contributions and interactions in pathophysiology remain unclear. ADHD undoubtedly reflects more than simply the amount of dopamine in a given synapse and will not only implicate autoreceptors and heteroreceptors but also likely involve other neurotransmitter systems (e.g., glutamate) and transcription factors controlling the way that brain systems are interwoven. Future studies will need to harness the power of QTL mapping and include gene-gene and gene-environment interactions to yield greater understanding of these complexities.

### *Autistic Disorder*

The initial molecular findings in autism predate those in other childhood-onset psychiatric disorders. Since the first description of whole blood hyperserotonemia in autism (Schain and Freedman 1961), numerous studies have identified elevated whole blood or platelet serotonin (5-HT) in more than 25% of patients with autism (Cook and Leventhal 1996). These findings led to consideration of the platelet as a model for the presynaptic neuron, since the platelet serotonin transporter takes up serotonin and the platelet 5-$HT_{2A}$ receptor binds to 5-HT released in the periphery. Increased platelet 5-HT transport and decreased platelet 5-$HT_{2A}$ binding were found in hyperserotonemic first-degree relatives of probands with autism compared with normoserotonemic first-degree relatives (Cook et al. 1993). Studies of central 5-HT or of its primary metabolite, 5-hydroxyindoleacetic acid (5-HIAA), assessed in the CSF have revealed no consistent abnormalities (reviewed in Cook 1990). In contrast, blunted response to fenfluramine (McBride et al. 1989) and worsening of stereotyped behavior after tryptophan depletion (McDougle et al. 1996) have been reported.

Although molecular imaging technology is difficult to apply in children with autism since informed consent or assent may be hard to obtain, investigators have started to study central 5-HT more directly. Two PET studies in autism using the radiolabeled 5-HT precursor α-methyl-L-tryptophan (AMT) found increased brain serotonin synthesis capacity in children over the age of 5 years in comparison to either epilepsy or sibling controls (Chugani et al. 1997, 1999). Furthermore, whereas the control children had increased 5-HT synthesis before the age of 5 that gradually decreased to adult levels, the children with autism had a steady increase to levels significantly higher than the adult control levels (Chugani et al. 1999). These studies have significant limitations, since only cross-sectional data were collected and appropriate pediatric age-, race-, and sex-matched controls could not be used because of ethical considerations of radiation exposure.

A number of studies have focused on other neuroendocrine or neurochemical markers. Dexamethasone nonsuppression, increased plasma β-endorphin, and increased plasma norepinephrine have all been reported. Two studies found decreased levels of plasma oxytocin in children with autism compared with control populations (Green et al. 2001; Modahl et al. 1998). A recent study reported increased blood spot levels of brain-derived neurotrophic factor (BDNF), neurotrophin 4/5, calcitonin gene–related peptide, and vasoactive intestinal peptide in neonates who later were diagnosed with autism or mental retardation (Nelson et al. 2001).

Immunological abnormalities have been reported in autism. The significance is unclear, and there is little to no evidence of typical autoimmunity in autism (reviewed in Korvatska et al. 2002). It is important to note that the development of the nervous system is highly dependent on molecular recognition to provide appropriate signals for development. Knockout of major histocompatibility class I function in mice led to non-immune-mediated facilitation of long-term potentiation and absence of long-term depression (Huh et al. 2000).

Molecular methodology is now being applied in postmortem studies of autism. These studies have tremendous potential, but initial findings may implicate a different distribution of neuronal types, numbers, and size, rather than implicating particular genes or protein systems in the pathophysiology of disease. Investigators have followed three approaches in examining areas of the brain with structural abnormalities in autism. The first approach is to consider important proteins within neurotransmitter systems. Two studies have considered the acetylcholine system and generated some interesting findings. One study used cortical autoradiography to demonstrate decreased radioligand binding to the muscarinic acetylcholine receptor $M_1$ and the nicotinic acetylcholine receptor subunits α4 and β2 (Perry et al. 2001). The nicotinic abnormalities were also seen in a comparison group with mental retardation. A follow-up study in the cerebellum used autoradiography to identify decreased radioligand binding to a nicotinic receptor subgroup, including subunits α3, α4, and β2 in all cell layers but increased radioligand binding to the α7 subunit in the granule cell layer (Lee et al. 2002). Immunohistochemistry confirmed these findings for the nicotinic receptor subunits α4 and α7. The meaning of these abnormalities in acetylcholine receptors remains unclear, but the importance of acetylcholine in learning and memory has been well documented.

Postmortem molecular studies have also been applied to other neurotransmitter systems. Autoradiography in the hippocampus revealed reduced radioligand binding to $GABA_A$ receptors but no change in binding to serotonin $5\text{-}HT_{1A}$ or serotonin $5\text{-}HT_{2A}$, muscarinic acetylcholine receptor $M_1$, glutamate NMDA, or kainate receptors, or high-affinity choline uptake sites (Blatt et al. 2001). One study used microarray technology to identify differing gene expression patterns in the cerebellum from people with autism and found increased mRNA for the genes encoding a glutamate transporter *(EAAT1)*, one of the glutamate AMPA receptors *(GLUR1)*, a GABA receptor *(GABRA5)*, and a glial protein *(GFAP)*, as well as changes in expression levels of a number of other genes (Purcell et al. 2001a). Notably, this study detected no change in the expression of the gene encoding glutamic acid decarboxylase, the enzyme that converts glutamate to GABA, although another study found decreased levels of this protein in cellular homogenates prepared from parietal and cerebellar regions (Fatemi et al. 2002). Follow-up studies of the genes identified with the microarrays found increased antibody binding to the EAAT1 and GluR1 proteins within cellular homogenates (Purcell et al. 2001a). In contrast to these findings, autoradiography showed that AMPA receptor density on the cell membrane was decreased in multiple cell layers in autism cerebellum (Purcell et al. 2001a).

The second approach to postmortem studies in autism, based on the finding of macrocephaly in a substantial minority of subjects with this disorder, is to consider proteins that control neuronal death. One research group has focused on expression of Bcl-2, a membrane-bound regulatory protein that inhibits programmed cell death (apoptosis). The initial study in homogenized cerebellar tissue found statistically significant decreases in anti-Bcl-2 antibody binding but only a nonsignificant trend for decreased neuronal β-tubulin antibody binding and essentially no difference in β-actin antibody binding (Fatemi et al. 2001a). A second study, using the same methodological approach, confirmed the decreased anti-Bcl-2 binding (Fatemi et al. 2001b). One study using parietal cell homogenates found no significant difference in anti-Bcl-2 binding; however, it found an increase in antibody binding to p53, a regulatory protein that favors apoptosis (Fatemi and Halt 2001).

The last approach is to consider peptides thought to be important in neuronal migration and synaptic development. The same group that studied Bcl-2 identified decreased antibody binding to reelin, an extracellular protein that is secreted by cerebellar granular cells and controls cell migration via cell surface receptor proteins (Fatemi et al. 2001b). Another research group performed a study in the cerebellum that found decreased antibody binding to only the long isoform of the neural cell adhesion molecule (NCAM), a cell surface protein thought to be important in brain development and synaptic plasticity. Unfortunately, this difference was not statistically significant after correction for the three NCAM protein isoforms, and mRNA levels for this protein were not significantly different (Purcell et al. 2001b). A study in the basal forebrain found increased antibody binding to BDNF, a signaling peptide important for synaptic development and function throughout the brain (Perry et al. 2001). Although these initial findings might be confirmed in other laboratories, immunohistochemistry may be necessary to accurately compare protein binding in relation to neuronal location and morphology.

Appropriate animal models are difficult to identify in autism, since the behavioral phenotype does not easily convert to lower mammals. The most appropriate models thus far have been in Rett syndrome and fragile X syndrome, both discussed earlier in the section on simple genetic disorders. Many transgenic mice demonstrate abnormalities in social behavior or behavioral inhibition in various experimental conditions, but it is difficult to equate these simple behaviors to the full autism phenotype. A few transgenic mouse lines have particularly interesting social phenotypes. Mice lacking *Dvl1,* a gene important in determining cell polarity, show abnormal sensory gating and reduced social interaction, including grooming of other mice and huddling and nesting behaviors (Lijam et al. 1997). Mice lacking the oxytocin gene *(Oxt)* fail to change their behavior on reexposure to a known cage mate, reflecting an apparent lack of social memory, despite normal performance on other olfactory and spatial memory tasks (Ferguson et al. 2000). On the other hand, mice that have had their arginine-vasopressin receptor 1A gene *(avpr1a)* promoter region replaced by prairie vole *avpr1a* promoter show increased affiliative behavior after injection of arginine-vasopressin (Young et al. 1999).

Human molecular genetics is a promising area of research in autism, since the recurrence rates among twins or siblings greatly exceed the rate in the general population. Evidence suggests that autism is a genetic disorder with complex inheritance. Recent estimates of the prevalence of autism with more complete ascertainment have been approximately 0.1%–0.2% (Chakrabarti and Fombonne 2001) for narrow diagnosis of autism and 0.6% for autism spectrum disorders. Twin studies show a 60%–91% concordance rate in monozygotic twins, depending on whether a narrow or a broad phenotype is considered, in contrast to no observations of concordance in dizygotic twins under narrow phenotypic definition and 10% concordance under broader phenotypic defintion (Bailey et al. 1995). The sibling recurrence rate has been estimated to be 4.5% (Jorde et al. 1991). The pattern of relative risk in autism is consistent with multiplicative inheritance, with multiple gene variants converging to lead to the phenotype.

Several chromosomal anomalies have increased rates in autism, including maternally inherited duplications of chromosome 15q11–q13, Angelman syndrome, Prader-Willi syndrome due to maternal uniparental disomy, FRAXA Down syndrome, Turner syndrome, and deletions of 2q37. None of these chromosomal disorders is found in more than 4% of autism samples. The most frequent chromosomal disorder in autism is probably maternally inherited duplication or triplication of 15q11–q13 (either interstitial duplications of 15q11–q13 or supernumerary inverted duplications) (Cook et al. 1997b; Schroer et al. 1998). This region includes at least two maternally expressed genes (*UBE3A* and *ATP10C*), but to date no variants in these genes have been identified that may account for other cases of autism. Several chromosomal anomalies have also been reported near regions of linkage on chromosomes 2q and 7q. Translocation breakpoint identification studies have identified two genes, *RAY1* and *AUTS2*, disrupted in single patients with autism, but abnormalities in the coding regions of these genes have not been found in a larger population of patients (Sultana et al. 2002; Vincent et al. 2000). Quite a number of other deletions or rearrangements have been reported scattered throughout the genome, leading some to speculate that families who have multiple offspring with autism may have a higher rate of errors in meiosis, perhaps as a result of abnormal chromatin structure (Yu et al. 2002).

Several candidate gene studies have been conducted on the basis of limited knowledge of neuropharmacology in autism, developmental neuropathological abnormalities, or chromosomal anomalies. The serotonin transporter gene *(SLC6A4)* is a candidate based on increased platelet serotonin uptake in a subgroup of hyperserotonemic (increased platelet serotonin) first-degree relatives of probands with autism and the responsiveness of OCD-related symptoms to potent serotonin transporter inhibitors (Cook et al. 1997a). 5-HTTLPR, a repeat polymorphism in the *SLC6A4* promoter, has been shown to affect transcription and is associated with neuroticism or anxiety in multiple samples (Lesch et al. 1996). In autism, most, but not all, studies have found nominally significant evidence of transmission disequilibrium at this polymorphism, but different alleles have been preferentially transmitted in different studies, suggesting either allelic heterogeneity or an as-yet-unidentified susceptibility variant. After a single point lod score of 3.6 was found at the intron 2 VNTR marker within *SLC6A4* (International Molecular Genetic Study of Autism Consortium 2001), a study was conducted to find other variants in *SLC6A4* in autism. This study tested transmission disequilibrium at single nucleotide polymorphisms (SNPs) throughout the *SLC6A4* gene region, finding a stronger signal at SNPs within intron 1A than at the original VNTR (Kim et al. 2002).

A number of groups have conducted family-based association studies of genes in the 15q11–q13 region implicated by chromosomal studies. A few transmission-disequilibrium studies of genes in this region, including *GABRB3*, have yielded nominally positive findings (Cook et al. 1998), but the findings from several others have been negative, as expected in a heterogeneous syndrome and when a marker rather than a functional variant is studied. Meta-analysis of all studies of the marker GABRB3 155CA-2 reveals a *P* value of 0.00022. A number of as yet unreplicated family-based associations have been noted in other genes in autism, including *HOXA1, AVPR1A, RELN,* and *WNT2.*

Several genomewide scans with relatively small sample sizes have been reported. Significant linkage findings have been reported at 2q (International Molecular Genetic Study of Autism Consortium 2001) and 3q (Auranen et al. 2002). None of these studies found definite evidence of linkage. Extension of the International Molecular Genetic Study of Autism Consortium sample has revealed multipoint lod scores over 3.0 for 2q, 7q, and 16p (International Molecular Genetic Study of Autism Consortium 2001). Interestingly, both the 2q and 7q linkage findings have strengthened in one or more samples when language impairment was added as a criterion for affection status (Buxbaum et al. 2001). Furthermore, the region on chromosome 7q was also implicated when age at first word was mapped as a QTL within autism sibling pairs (Alarcón et al. 2002). It is also of interest that a study of ADHD found significant linkage at 16p, overlapping with the linkage finding in autism in the same region (Smalley et al. 2002).

Aside from study of general language impairment, there have been relatively few attempts to look at subgroups or genotype-phenotype correlation in autism. Macrocephaly, language level, general level of intellectual functioning, blood (platelet) serotonin levels, degree of social and communication impairment, presence of seizure disorder, gender of proband, dysmorphology, and severity or specific nature of restricted and repetitive behaviors are phenotypes of particular interest. An alternative approach might be to map loci for some of the traits that cluster in families of a probands with autism.

## *Tourette Syndrome, Tic Disorders, and Childhood-Onset Obsessive-Compulsive Disorder*

Gilles de la Tourette syndrome (GTS) is a syndrome with a prevalence rate of around 0.05%, with more affected males. GTS, other tic disorders, and symptoms or full expression of childhood-onset OCD are associated both within individuals and within families. Molecular approaches have overlapped among these disorders, particularly in immunology. The three conditions will be considered together here, but an independent review of OCD in adults may be helpful for the interested reader (Pato et al. 2001). Molecular investigation in GTS, tic disorders, and OCD has focused on neurochemical, immunologic, and genetic approaches.

Since dopamine-blocking drugs reduce tics and dopaminergic drugs increase them, initial molecular studies focused on both central and peripheral assessment of the dopaminergic system in GTS. Multiple studies were unable to demonstrate consistent perturbation in central or peripheral dopamine, but varied methodology was applied across studies (Anderson et al. 1999). More recent studies have considered specific proteins within the dopaminergic system but have also been met with equivocal results. Some, but not all, recent studies have identified increased binding to the dopamine transporter protein both postmortem (Singer et al. 1991) and in vivo (Heinz et al. 1998; Malison et al. 1995; Muller-Vahl et al. 2000a). An increased in vivo binding to the dopamine $D_2$ receptor reported by Wolf and colleagues (1996) has not been identified in other studies (George et al. 1994; Muller-Vahl et al. 2000b; Turjanski et al. 1994). Due to obvious technological limitations, all in vivo imaging studies are unable to differentiate between altered protein expression in the context of normal neuronal number or structure, or vice versa, or alterations in both. Quantitative measures addressing self-injurious behavior, impulse control, and severity of vocal tics have been applied in some studies (Muller-Vahl et al. 2000a; Wong et al. 1997), but while these approaches are likely to be helpful to parse out heterogeneity within the syndrome, much larger sample sizes will be necessary.

Neurochemical investigation has also focused on norepinephrine (NE) and serotonin (5-HT), as well as scattered studies of other neurotransmitters. A GTS study of CSF NE and its metabolite MHPG (3-methoxy-4-hydroxyphenylglycol) demonstrated increased levels that some have attributed to a heightened response to the stress of lumbar puncture (Leckman et al. 1995). In one study, patients with GTS also had elevated CSF levels of corticotropin-releasing hormone (CRH), another component of the stress response (Chappell et al. 1996). The findings of lowered GTS plasma levels of tryptophan (Comings 1990; Leckman et al. 1984), the precursor amino acid of 5-HT, agree with findings of reduced tryptophan, 5-HT, and its metabolite 5-hydroxyindoleacetic acid (5-HIAA) in postmortem brains (Anderson et al. 1992). Studies in childhood-onset OCD find no overall change in whole-blood 5-HT levels, but a positive family history predicted higher levels (Hanna et al. 1991). Some, but not all, studies report decreased binding to the platelet serotonin transporter protein in pediatric-onset OCD compared with controls (Sallee et al. 1996; Stein and Uhde 1995). One postmortem study in GTS reported reduced glutamate in the medial globus pallidus (Anderson et al. 1992), but this study remains to be replicated. An in vivo study in childhood OCD revealed elevated glutamatergic MRS signal in the left caudate of patients that reverted to control levels after treatment with paroxetine, correlating with change in symptom severity (Rosenberg et al. 2000). Interested readers may find helpful a recent review of neurochemistry and neurobiology in adult and childhood-onset OCD (Stein 2000).

A number of mouse models have been employed in translational research into the origins of GTS or OCD. Pharmacological models demonstrated that injections of dopaminergic or glutamatergic agents into the striatum induced rapid, repetitive head and forelimb movements (Karler et al. 1995). More recent models have used transgenic animals. In one model, investigators attached a portion of the cholera toxin gene to the dopamine $D_1$ receptor promoter (Campbell et al. 1999). A number of lineages were generated that had a variety of regional expression patterns of the toxin, which increases cAMP levels and thereby potentiates the affected neurons. One of the mouse lines expressed the toxin in glutamatergic cortical neurons and the intercalated nucleus of the amygdala and exhibited repetitive behaviors, including stereotypic repetition of normal movements, grooming behaviors, and leaping, all of which worsened with glutamatergic but not dopaminergic agents (McGrath et al. 2000). In a potential extension of the model to include tic disorders, these mice, particularly males, exhibit sudden jerks of the head or body that improve with the norepinephrine receptor $\alpha_2$ autoreceptor agonist clonidine (Nordstrom and Burton 2002).

A second animal model resulted from knockout of the *Hoxb8* gene, one of a group of "homeobox-containing" transcription factor genes that determine the morphology of different body segments. These knockout mice show excessive grooming behavior that leaves portions of their bodies bare or even bleeding, much as in trichotillomania (Greer and Capecchi 2002). As these models and the understanding of the molecular basis of the human disorders start to converge, investigators may be able to develop and test novel treatments in animals before application to humans.

Much excitement has been generated by immunology research in tic disorders and childhood-onset OCD. A new entity, pediatric autoimmune neuropsychiatric disorders associated with streptococcal infections (PANDAS), has even been proposed to describe an autoimmune etiology in a small subpopulation of patients with these disorders (Allen et al. 1995). The original hypothesis arose from two observations. First, some patients with rheumatic fever, one of the autoimmune sequelae of streptococcal infection, develop Sydenham's chorea, a movement disorder that often includes OCD symptoms (Swedo 1994). Second, some patients with OCD or tic disorders appear to develop the disorder or exacerbations in temporal relationship to streptococcal infections (see Swedo et al. 1997), although this has not been established in a prospective study. First-degree relatives of patients with PANDAS also show elevated rates of OCD and tic disorders (Lougee et al. 2000). An initial controlled study found that plasma exchange or intravenous immunoglobulin improved symptoms for a select group of patients with PANDAS (Perlmutter et al. 1999). An open trial did not support plasma exchange in a small group of children with childhood-onset OCD without poststreptococcal exacerbations (Nicolson et al. 2000). Neither therapy has been compared with standard pharmacological or psychotherapeutic approaches.

Molecular techniques have been used to explore the relationship between autoimmunity and tic disorders or OCD, both in humans and in lower mammals. The monoclonal antibody D8/17 has been studied as a potential marker for poststreptococcal autoimmune susceptibility or disease. Zabriskie and colleagues (1985) originally identified D8/17 as a marker that binds to a larger population of B cells in patients with rheumatic fever compared with controls. Two initial studies performed in the Zabriskie laboratory found increased D8/17 binding both in a PANDAS population and in a general population of patients with childhood-onset OCD or tic disorder, including GTS (Murphy et al. 1997; Swedo et al. 1997). Unfortunately, methodological difficulties have plagued attempts to replicate these findings. Initial D8/17 findings were based on qualitative "positive" versus "negative" analyses that assumed a bimodal distribution and required a technician to count each individual cell. This subjective approach showed too much variability to be generally useful across laboratories. The actual output of the assay, however, is in percentage of positive B cells, a quantitative measure, and subsequent studies have incorporated an objective approach, flow cytometry, to demonstrate that D8/17 binding follows a continuous rather than a bimodal distribution. Reports using objective, quantitative methodology are starting to provide preliminary support for the initial findings, but a standard methodology has yet to emerge to enable replication (Chapman et al. 1998; Hoekstra et al. 2001; Murphy et al. 2001). If further studies confirm the importance of this marker, further investigation may identify the B cell antigen itself and explore its significance, whether as a marker of susceptibility or of disease. It is notable that increased D8/17 reactivity has also been reported in autism, anorexia nervosa, and susceptibility to rheumatic fever.

Although markers like D8/17 may lend support for the importance of autoimmunity in tic disorders and OCD, they do not allow a direct understanding of pathophysiology. Some investigators have focused on auto-antibodies that could theoretically contribute to disease. A few studies have identified increased antineuronal antibodies in children with tic disorders (Kiessling et al. 1993; Laurino et al. 1997; Singer et al. 1999). Since the pathophysiology of OCD and tic disorders is thought to specifically involve the basal ganglia (Rapoport and Fiske 1998), investigators have been excited to find specific antibodies to the striatum in GTS (Hallett et al. 2000; Kiessling et al. 1993; Singer et al. 1998; Wendlandt et al. 2001); however, the mere presence of auto-antibodies does not prove their involvement in pathophysiology, nor does it prove an autoimmune contribution to pathophysiology (Murphy and Goodman 2002). The patients with GTS had more auto-antibodies, but the control subjects also demonstrated auto-antibodies, suggesting that their presence does not necessarily indicate pathology. Two recent studies demonstrated that auto-antibodies from the sera of GTS patients caused increased vocalizations and repetitive movements when introduced directly into the striatum of rats (Hallett et al. 2000; Taylor et al. 2002). A number of gaps need to be clarified before this could represent a plausible pathophysiology. For example, how would these auto-antibodies cross the blood-brain barrier? Further, to what cellular elements are the antibodies binding? Interestingly, infusion of dopaminergic agents in the same region in the rat produces similar behavioral change (Bordi and Meller 1989; Delfs and Kelley 1990).

Molecular genetic research in GTS and childhood-onset OCD initially followed the candidate gene approach but is now incorporating model-free, genomewide linkage and association methods. There is significant evidence for a genetic component in GTS. Twin studies have found a monozygotic concordance rate of 53%–56% and a dizygotic concordance rate of 8% (Hyde et al. 1992; Price et al. 1985). This is consistent with a reported first-degree relative recurrence rate of around 9%. The recurrence rate is higher among male relatives. When affection status is broadened to include chronic tics or OCD, a much higher recurrence rate is observed. On the other hand, the relative recurrence risk in childhood-onset OCD is less well studied. Family studies using adult probands have found that an early age at onset of obsessive-compulsive symptoms is strongly related to a more familial form of the disorder (Nestadt et al. 2000; Pauls et al. 1995). Family studies involving child and adolescent probands have also consistently indicated that OCD is familial (Lenane et al. 1990; Leonard et al. 1992). The rate of OCD in the first-degree relatives of the probands in those studies has ranged from 8% to 17%. The familial overlap between GTS, tic disorders, and OCD provides a challenge in genetic studies of each disorder. The correct approach will likely be identified not by theory but by results once chromosomal regions and genes are identified.

A number of chromosomal findings have been reported in GTS. A balanced translocation that segregates with GTS was reported in several family members between chromosomes 7q22–31 and 18q22.1 (Boghosian-Sell et al. 1996). Further interest has centered on chromosomes 7 and 18. Deletion of chromosome 18q22.1 was observed in one patient with GTS (Donnai 1987). Inverted duplication of chromosome 7q22.1–31.1 was found in one patient with GTS and dysmorphic features. Both the breakpoint and the insertion site of this inverted duplication disrupted *IMMP2L*, the inner mitochondrial membrane peptidase subunit 2–like gene (Petek et al. 2001). This disruption leaves only one copy of the *IMMP2L* gene on the normal chromosome. Reported genetic imprinting on 7q31 could complicate a dosage effect of gene expression in this region. A number of other genes in the region are duplicated and could also be implicated.

Candidate gene studies in GTS have focused on the dopaminergic system without consistent success. One family-based association study reported transmission disequilibrium at the dopamine receptor $D_4$ gene *(DRD4)* (Grice et al. 1996), but this finding has not yet been replicated in subsequent studies.

Linkage analysis in GTS has been less fruitful than was initially expected. Despite a number of focused linkage studies, no evidence has been found for linkage to the regions implicated by chromosomal findings. Nonparametric linkage analysis using affected sibling pairs also failed to find significant evidence for linkage, but this approach identified two regions with suggestive linkage on chromosomes 4q and 8p (The Tourette Syndrome Association International Consortium for Genetics 1999). One group applied genomewide case-control association methods in the Afrikaner population (Simonic et al. 1998). Although none of these regions appear to reach stringent genomewide significance by criteria suggested for association studies, a follow-up family-based association study in an independent Afrikaner population replicated association on chromosomes 2p11, 8q22, and 11q23–24 (Simonic et al. 2001). Furthermore, a linkage study considering only markers associated in the Afrikaner population found significant linkage on chromosome 11q23 in a single large French Canadian pedigree (Merette et al. 2000).

A small genome screen in extended families ascertained through probands with pediatric-onset OCD found suggestive linkage on chromosome 9p (Hanna et al. 2002) , which coincides with a chromosomal deletion observed in GTS (Taylor et al. 1991). A linkage study in a group of families with GTS probands harnessed the overlap between GTS and OCD to interesting effect. Based on factor analysis of OCD symptoms (Leckman et al. 1997), hoarding behavior was considered as a quantitative trait, with resulting significant linkage on chromosomes 5q and 17q, as well as on chromosome 4q, near the suggestive linkage finding in GTS (Zhang et al. 2002). When hoarding was considered as a dichotomous trait, no significant evidence for linkage was detected. The candidate gene approach has not generated significant association findings in childhood-onset OCD; although association studies have implicated a number of genes in adult-onset OCD, including *HTR1B, COMT, HTT,* and *MAOA,* that have not yet been examined in childhood-onset samples.

## *Eating Disorders*

Central nervous system neurochemistry and genetics have dominated research on the molecular neurobiology of eating disorders. Thus far, neurochemistry studies have yielded the most promising findings, but genetic findings are beginning to emerge as well. Research in anorexia nervosa (AN) and bulimia nervosa (BN) confronts dramatically the issue of state versus trait. Particularly in AN, when patients in the full grips of the disorder are severely malnourished, disruptions in neurochemical and neuropeptide systems are likely to mark the starvation state rather than the underlying neurobiology of the disorder. Investigators must study individuals both in the grips of disease and in remission to identify trait markers that put them at risk of the disease (Kaye et al. 2000).

A number of neurohormonal and neuropeptide systems appear to be disrupted during the active phase of the disease but not during remission; these systems include the stress hormones cortisol (Walsh et al. 1978) and CRH (Gold et al. 1986), as well as neuropeptide/neurohormones, including leptin (Hebebrand et al. 1995; Mantzoros et al. 1997), oxytocin (Demitrack et al. 1990), and neuropeptide Y (Kaye et al. 1990). Other studies have found abnormalities in the remitted state that have not been studied in the active state: reduced CSF gastrin-releasing peptide (GRP) levels in women with a history of BN (Frank et al. 2001a) and reduced CSF galanin (GAL) levels in women with a history of AN (Frank et al. 2001b). Arginine-vasopressin levels are low in the CSF of patients with active AN (Gold et al. 1983) but elevated in the CSF of patients with recovered AN, bulimic subtype (Frank et al. 2000).

Abnormalities in the serotonergic system may characterize both the ill and the recovered state. These abnormalities are made even more interesting by the fact that dietary changes can acutely modify the peripheral concentration of the amino acid tryptophan, the precursor to 5-HT. Patients with active AN have decreased concentrations of the 5-HT metabolite 5-HIAA in their CSF (Kaye et al. 1998), while recovered AN patients have increased concentrations of CSF 5-HIAA (Kaye et al. 1991). Patients with active BN have blunted neuroendocrine responses to 5-HT agonists (Brewerton et al. 1992; Jimerson et al. 1997), while recovered BN patients have increased levels of CSF 5-HIAA (Kaye et al. 1998). SPECT imaging has revealed decreased 5-HT transporter binding in patients with BN or binge eating during their illness (see Kuikka et al. 2001; Tauscher et al. 2001). In women who have recovered from bulimia nervosa, PET imaging demonstrated decreased serotonin$_{2A}$ (5-HT$_{2A}$) binding in the medial orbital frontal cortex, as well as a lack of the normal age-related decline in 5-HT$_{2A}$ binding (Kaye et al. 2001).

Anorexia nervosa and bulimia nervosa appear to have some overlapping genetic susceptibility. AN has an incidence of about 0.5% among women and has the highest mortality rate among psychiatric disorders. BN has an incidence of about 1.5% among women, but it is much less likely to be lethal. The rates in men are approximately one-tenth of the rates in women. The largest twin study in AN found a monozygotic concordance rate of 56% in comparison to 5% among dizygotic twins (Holland et al. 1988). Twin studies in BN report monozygotic concordance rates ranging from 8% to 23% compared with 0 to 9% in dizygotic twins (Bulik et al. 2000). A recent large family study used both AN and BN probands to demonstrate a significant overlap in genetic susceptibility. The rate of AN was 3.4% in first-degree relatives of AN probands. The rate of BN was 4% in first-degree relatives of BN probands. Interestingly, the rates of each disorder in the relatives of probands with the other disorder were almost identical (Strober et al. 2000).

Initial genetic studies in eating disorders focused on candidate genes. In anorexia nervosa, groups have focused particularly on the serotonergic, dopaminergic, and leptinergic systems without much success. Much interest has centered on the serotonin 5-$HT_{2A}$ receptor gene, but a large family-based study revealed no evidence of association at this gene (Gorwood et al. 2002). Single family-based studies in AN have found associations at a polymorphism within the norepinephrine transporter gene *(NET)* promoter region (Urwin et al. 2002) and with a high-functioning allele of the catechol-*O*-methyltransferase gene *(COMT)* (Frisch et al. 2001). This latter association is somewhat unexpected, since the opposite allele is associated with OCD, which occurs at an increased rate in AN probands and their relatives (Lilenfeld et al. 1998). Another family-based study found significant association with long alleles of a triplet repeat polymorphism in the neuronal potassium channel gene, *KCNN3* (Koronyo-Hamaoui et al. 2002).

A collaborative linkage study recently reported initial results, including suggestive linkage to a region on chromosome 1p in families with two or more relatives with restricting AN, a subtype lacking any binging or purging behavior (Grice et al. 2002). Linkage analysis incorporating concordance for two quantitative traits, drive for thinness and obsessionality, identified three different linkage peaks: chromosome 1q, with a lod score nearing genomewide significance, and suggestive evidence on chromosomes 2p and 13q (Devlin et al. 2002). Linkage analysis that considered the entire group of subjects affected with AN, BN, or eating disorder NOS (not otherwise specified) revealed no suggestive evidence for linkage. Larger sample size in the various individual eating disorders and in the overall sample may help clarify the relationship between AN, BN, and related eating disorders.

## *Reading Disability*

The molecular neurobiology of reading disability is largely limited to molecular genetics at this point; however, linkage mapping of this disorder is already quite advanced. Susceptibility to developmental dyslexia, or specific reading disability, has a substantial genetic component. Dyslexia is a common disorder, with a prevalence of 5%–10%, with a male-female ratio as high as 4:1. The largest twin study found a concordance rate of 68% in monozygotic twins, in comparison to 38% in dizygotic twins. Family studies have found a recurrence rate of around 35%–45% in first-degree relatives.

A number of chromosomal regions have been implicated by linkage findings in dyslexia. The most evidence points to a 5–10 cM region on chromosome 6p21.3. A QTL for a composite measure of reading disability was first mapped near chromosome 6p21 (Cardon et al. 1994). Subsequent studies have narrowed the 6p QTL to include some components of reading disability (orthographic coding and phonological decoding) more than other components (phoneme awareness and word recognition) (Fisher et al. 1999; Gayan et al. 1999; Grigorenko et al. 1997). A recent study found both linkage and association, narrowing the most likely region of interest to four mega-bases on chromosome 6p22.3 (Kaplan et al. 2002). Notably, this locus does not affect IQ, but one recent linkage study within reading disability sibling pairs revealed a weak linkage peak for comorbid ADHD in this region even when reading disability itself was regressed out (Willcutt et al. 2002). The genetic defect responsible for this well-replicated linkage finding has yet to be identified.

A few other regions of linkage have also been identified in dyslexia. A region on chromosome 15q21 has been implicated in multiple linkage studies (Grigorenko et al. 1997), and a balanced translocation involving chromosome 15q21–22 has been reported to segregate with dyslexia in one family (Nopola-Hemmi et al. 2000). A recent family-based study found significant association with chromosome 15q15 markers in two groups of probands (Morris et al. 2000). Likewise, linkage with a region on chromosome 1p34–36 has been found in multiple studies (Grigorenko et al. 2001), and a balanced translocation of chromosomes 1p22 and 2q31 was reported in a family with dyslexia and speech delay (Froster et al. 1993). Most recently, a locus at 18p11.2 has been reported for dyslexia in independent samples reported in a single paper (Fisher et al. 2002a). Regions of linkage have also been identified in single samples, including pericentromeric chromosome 3 and chromosome 6q, but these have not yet been confirmed in an independent sample.

## *Communication Disorders*

Molecular research in communication disorders is also limited primarily to genetic approaches. Communication disorders can be divided into two types: speech disorders and language disorders. Stuttering is a disorder of speech fluency. Twin and family studies have shown that both persistence and recovery from stuttering have heritable components (Yairi et al. 1996). A large sample of sibling pairs concordant for stuttering is being collected for linkage analysis (N. J. Cox et al., personal communication, 2002).

*Specific language impairment* refers to a developmental disorder leading to language deficits in children with otherwise intact cognition. Twin studies suggest that specific language impairment has a significant genetic component as well; however, most families with language disorders show complex patterns of inheritance. Two linkage studies have identified regions of interest in this disorder. One study identified a region with significant evidence for linkage on chromosome 13q21 in five Celtic families with specific language impairment (Bartlett et al. 2002). Linkage analysis also yielded significant evidence for linkage in a single four-generation extended family with an autosomal dominant speech and language disorder, orofacial dyspraxia, and deficits in language processing and grammatical skills, despite relatively spared nonverbal intelligence. The original linkage finding, together with translocation-mapping in an unrelated patient, allowed identification of the first speech and language disorder susceptibility gene on chromosome 7q31, *FOXP2,* which encodes a putative transcription factor (Lai et al. 2001). The *FOXP2* mutation has not been found in subsequent studies of language impairment or autism (Meaburn et al. 2002; Newbury et al. 2002; Wassink et al. 2002). A recent study demonstrated an uncommon pattern of a high level of divergence between chimpanzees and humans relative to high conservation of the gene between mice and humans (Enard et al. 2002).

## Complex Genetic Disorders With Variable-Onset Psychiatric Phenotypes

### *Anxiety Disorders and Behavioral Inhibition*

Childhood anxiety disorders do not necessarily relate directly to their adult counterparts, complicating any analysis of their genetics. For example, childhood generalized anxiety disorder (GAD) appears to put patients at risk of not only adult GAD but also other anxiety disorders and depression. Other disorders, such as social phobia and blood-injury phobia, commonly have onset in childhood or adolescence and persist into adulthood, but the relationship between age at onset and degree of familiality is not well studied.

Despite their quite different manifestations, longitudinal and family studies have made a connection between childhood separation anxiety disorder and panic disorder in adulthood. Furthermore, recent work suggests that children with separation anxiety disorder, and to a lesser extent other anxiety disorders, react to increased carbon dioxide in much the same way as adults with panic disorder (Pine et al. 1998, 2000). One group found that childhood separation anxiety disorder was associated with a higher familial loading of panic disorder (see Battaglia et al. 1995). Molecular studies have not yet focused on childhood anxiety disorders or their relationship to adult disorders.

Since childhood anxiety disorders appear at least somewhat discontinuous with adult disorders, a more useful approach may be to consider behavioral measures of anxiety as a trait that underlies both adult and childhood-onset anxiety disorders. Behavioral inhibition to the unfamiliar (BI) has been studied as such an observable trait that can be evaluated in the laboratory in children as young as 14 months (Kagan et al. 1988). This trait appears to be stable and heritable and to be associated with risk of anxiety disorders in children with BI and their family members. As a directly observed measure during early childhood, it may be less encumbered with environmental effects than psychiatric assessments in adults. It also corresponds nicely to similar measures in animals. An initial study targeted four candidate genes identified through mouse models without finding significant association after statistical correction (Smoller et al. 2001), but this approach has much promise.

### *Childhood-Onset Conduct Disorder (Early-Onset Antisocial Behavior)*

A great deal of societal interest in antisocial behavior and aggression has led to some investigation into the molecular bases of early-onset antisocial behavior as defined by oppositional defiant disorder (ODD) and conduct disorder (CD). Despite the diagnostic distinction between ODD and CD, recent studies suggest common genetic liability and longitudinal overlap between the two disorders (Eaves et al. 2000; Lahey et al. 1999). Analysis of genetic contributions to conduct disorder symptoms is also complicated by the increase in these symptoms seen in more recent age cohorts (Jacobson et al. 2000). Furthermore, studying conduct disorder as an isolated entity is quite difficult, since comorbidity with other disorders is so common. In particular, conduct disorder is much more common in patients with ADHD. Despite all of these complicating factors and disparate findings across studies, a consensus has emerged that the interplay between genetic and environmental factors makes a significant contribution to early-onset aggression and conduct disorder (Simonoff 2001).

A few neurochemical and endocrine studies have emerged in early-onset antisocial behavior. These studies have largely attempted to extend findings from adult antisocial or criminal populations, including decreased CSF levels of the serotonin metabolite 5-HIAA and decreased cortisol stress response. A study in males with conduct disorder found significant correlations between early onset and severity and elevated whole blood serotonin levels (Unis et al. 1997). Measurement of whole blood serotonin in a birth cohort study found elevated levels in males with a history of violence before the age of 21 (Moffitt et al. 1998). Elevated prolactin response to the serotonergic drug fenfluramine also correlated with measures of aggression in a small group of brothers of juvenile offenders (Pine et al. 1997). Lower baseline serum or salivary cortisol levels in children or adolescents with conduct disorder or aggressive behavior have been reported (McBurnett et al. 2000; Pajer et al. 2001; Schulz et al. 1997). This is a remarkable finding given that these children typically are exposed to more psychosocial stressors than are controls. Some, but not all, studies have identified increased serum levels of androgens, particularly dehydroepiandrosterone sulfate (DHEAS), an adrenal androgen, in children with conduct disorder (Dmitrieva et al. 2001; Scerbo and Kolko 1994; van Goozen et al. 1998). Matching subjects with appropriate controls may be difficult in the child and adolescent population, since androgen levels are determined not just by age but also by pubertal stage.

Quite a few animal studies have considered aggressive behavior as a potential phenotype. Mice lacking the gene encoding monoamine oxidase A *(MAOA),* which degrades serotonin and norepinephrine, exhibit dramatically increased aggression (Cases et al. 1995). Increased aggression also occurs in a number of other transgenic mouse lines, including mice lacking either serotonin$_{1B}$ (5-HT$_{1B}$) receptors (Saudou et al. 1994); α-CaMKII, a calmodulin kinase (Chen et al. 1994); or neuronal nitric oxide synthase (Nelson et al. 1995). It is interesting to note that mice lacking the serotonin transporter gene *(htt)* exhibit reduced levels of aggression (A. Holmes et al. 2002). Young, but not old, knockout mice lacking the long form of the dopamine $D_2$ receptor ($D_2$L) show lower levels of aggression when intruder mice are introduced into the home cage (Vukhac et al. 2001), a potential model for age-dependent aggression. Animal studies also allow manipulations of the environment to evaluate potential gene-environment interactions. A study in rhesus monkeys showed that the serotonin transporter promoter polymorphism has an effect on CSF 5-HIAA only when coupled with adverse rearing (Bennett et al. 2002).

Molecular genetic studies have not yet tackled the complexities of early-onset antisocial behavior in humans. As mentioned in the ADHD section, comorbidity between ADHD and conduct disorder may identify a population of children with a higher genetic loading than for either disorder alone. One family-based study found association between the 7-copy of the DRD4 gene in children with comorbid ADHD and conduct disorder, despite a negative finding for ADHD alone (J. Holmes et al. 2002). A clever analysis of a longitudinal study found that an interaction between childhood mistreatment and a low-activity version of the *MAOA* promoter predicted antisocial behavior in adulthood (Caspi et al. 2002), although maltreatment itself was a more significant predictor. Coupled with earlier work demonstrating serotonergic abnormalities in both lower primates (Coplan et al. 1998) and human children (Pine et al. 1997) raised in adverse environments, this study launches a new era of research on gene-environment interactions in psychiatry.

## Mood Disorders

Early-onset bipolar disorder can be divided into two types: adolescent-onset and prepubertal-onset. The adolescent-onset form of the disorder tends to be more severe than later-onset bipolar illness, and family studies report a higher recurrence rate in relatives of adolescent-onset probands than in those of later-onset probands (Schurhoff et al. 2000; Strober et al. 1988). The nosology and epidemiology of prepubertal bipolar disorder have been controversial, in part because of symptom overlap and possible comorbidity with ADHD. Therefore, research diagnosis explicitly requires symptoms other than hyperactivity and irritability, such as grandiose behavior, decreased need for sleep, racing thoughts, and elated mood, and these symptoms have been demonstrated to be more severe in a group of children with childhood-onset bipolar disorder relative to ADHD and normal control groups (Geller et al. 2002). Small family studies in prepubertal-onset bipolar disorder reveal elevated rates of the disorder among first-degree relatives compared with the rates seen in adult-onset probands (Todd et al. 1993; Wozniak et al. 1995). ADHD is also diagnosed at higher rates in relatives of adult probands with bipolar disorder (Faraone et al. 1997). The relatively recent recognition of prepubertal bipolar disorder means that some children initially diagnosed with ADHD may eventually be diagnosed with bipolar disorder.

Early onset of major depressive disorder is linked with delayed recovery, high risk of recurrent depression, and increased comorbidity with other disorders, including substance abuse (Alpert et al. 1999). Increased severity suggests that molecular abnormalities may be more marked in the early-onset form of the disorder, but unfortunately this has not been the case thus far. Investigators have used peripheral endocrine and neurochemical challenge measures to explore underlying abnormalities in childhood or adolescent major depressive disorder. Blunted growth hormone response to various stimuli has been identified in multiple studies of both children with depression and those at high genetic risk (Birmaher et al. 2000; Ryan et al. 1994). Other peripheral molecular findings have been less consistent. Some studies show disturbances in the hypothalamic-pituitary-adrenal (HPA) axis in depressed children, but even the dexamethasone suppression test has shown only inconsistent abnormalities (Birmaher et al. 1996; Ryan et al. 1992). One group has found decreased cortisol response and increased prolactin response (in females only) to the serotonin precursor 5-hydroxy-L-tryptophan (L-5HTP) in both depressed children and children at high risk of depression (Birmaher et al. 1997). These initial findings will hopefully become more specific as more direct measures of these systems emerge.

While some controversy remains, in most studies early onset of major depressive disorder predicts higher familial loading. Conversely, family history predicts risk of recurrent depression in a child initially presenting with major depressive disorder (Wickramaratne et al. 2000). These observations suggest sibling pairs with early-onset major depressive disorder and a positive family history as the ideal population for initial linkage and association studies.

Genetic findings in early-onset major depressive disorder have not yet emerged. Initial candidate gene studies in prepubertal and early-adolescent bipolar disorder have been negative, but genome-wide linkage studies, which may be more powerful than adult-onset bipolar disorder studies as for other early-onset samples, have not yet been conducted.

## Summary

The dominant theme in the molecular biology of child psychiatry is complexity (see Table 1–1 for a summary of molecular genetic and cytogenetic findings in psychiatric disorders with onset in childhood or adolescence). Even relatively simple genetic disorders like Rett syndrome have allelic heterogeneity leading to different phenotypes depending on the mutation. Syndromes like autism are likely to show a large degree of both genetic and neurobiological heterogeneity, with defects in multiple brain regions or neuronal types and variants in many different gene combinations producing clinically indistinguishable phenotypes, at least at this point. Some disorders, like attention-deficit/hyperactivity disorder, may be the summation of multiple quantitative trait loci that produce a spectrum with maladaptive behavior at its extreme.

It is sometimes assumed that the study of molecular abnormalities discounts the impact of environment on disease. However, as we learn to sort out the complex relationships between genetic variants, molecular neurobiology, and disease, we must also start considering the relationship between genetic and environmental influences. With advancing molecular genetic methods, we may be able to identify all of the susceptibility variants for a given child psychiatric disease; however, our mission does not stop there. Studies of protein systems may identify environmental perturbations that impact disease course or clinical picture. While it is possible to identify environmental factors in multifactorial diseases without any genetic knowledge, this task will become much easier when the genetic susceptibility of each individual can also be understood. The intersection of genetic and environmental studies may yield unsuspected maternal, placental, and childhood environmental factors that increase sus-

**Table 1–1.** Molecular genetic and cytogenetic findings in psychiatric disorders with onset in childhood or adolescence

| Disorder | Chromosomal anomalies | Significant or replicated linkage | Replicated family-based association[a] |
|---|---|---|---|
| Attention-deficit/hyperactivity disorder | | 16p13 | *SLC6A3=DAT*; *DRD4*; *DRD5*; *SNAP-25* |
| Autistic disorder | 15q11–13 duplications; multiple other anomalies | 2q31; 3q26; 7q21–33; 16p13 | *SLC6A4=HTT*; *GABRB3* |
| Tourette syndrome | 7q22–31:18q22 translocation; other 7q and 18q anomalies | 11q23 | |
| Hoarding phenotype in Tourette syndrome[b] | | 5q35; 17q25; 4q34–35 | |
| Anorexia nervosa | | 1q31[c] | |
| Dyslexia (QTL analyses) | | 6p21; 15q21; 1p34–36; 18p11 | |
| Specific language impairment | | 7q31[d]; 13q21 | |

*Note.* QTL=quantitative trait locus.
[a]Replicated in at least one but not all studies.
[b]Surrogate for subtype of obsessive-compulsive disorder.
[c]Covariates of Obsessionality and Drive for Thinness.
[d]Specific mutations in *FOXP2* identified in a three-generation pedigree (known as the KE family) with language impairment.

ceptibility to disease or worsen disease course.

While we are hopeful that molecular methods will eventually reap vast rewards for children and families struggling with childhood- and adolescent-onset psychiatric disorders, more resources clearly need to be dedicated toward this goal. Twenty-eight percent of the U.S. population is under 20 years old (U.S. Census Bureau 2001), but a comparison with the articles detailing adult disorders in this volume will reveal that a far lower percentage of molecular data has been generated in child and adolescent psychiatry. This does not mean that funding for adult disorders should be diminished, since these disorders are still poorly funded relative to their societal impact, but research funding and career development opportunities need to be greatly expanded in child psychiatry.

If the promise of molecular neurobiology is realized in child psychiatry, we will gain not only greater understanding but improved environmental and pharmacological treatments. As pathophysiology is determined to be related to novel protein systems, new avenues for drug development will yield treatments that may address a susceptibility factor at a particular time point or within a particular environmental context. Prevention of disease may become possible as early identification and efficacious treatments converge. In addition, identification of protective factors may guide treatment or prevention by the development of interventions analogous to endogenous processes. The complexity of psychiatric disease makes the task appear daunting, but when multiple factors act in concert to cause disease, it may be possible to prevent or treat disease by correcting only one of them. A potent example of this phenomenon is the insulin locus in type I diabetes, which generates only a weak genetic association but corresponds to a life-saving treatment. A similar discovery in child psychiatry may seem likely only in the remote future, but it may already be lurking in early molecular reports.

## References

Alarcón M, Cantor R, Liu J, et al: Evidence for a language quantitative trait locus on chromosome 7q in multiplex autism families. Am J Hum Genet 70:60–71, 2002

Allen AJ, Leonard HL, Swedo SE: Case study: a new infection-triggered, autoimmune subtype of pediatric OCD and Tourette's syndrome. J Am Acad Child Adolesc Psychiatry 34:307–311, 1995

Alpert JE, Fava M, Uebelacker LA, et al: Patterns of Axis I comorbidity in early-onset versus late-onset major depressive disorder. Biol Psychiatry 46:202–211, 1999

Amir RE, Van den Veyver IB, Wan M, et al: Rett syndrome is caused by mutations in X-linked MECP2, encoding methyl-CpG-binding protein 2. Nat Genet 23:185–188, 1999

Anderson GM, Pollak ES, Chatterjee D, et al: Postmortem analysis of subcortical monoamines and amino acids in Tourette syndrome. Adv Neurol 58:123–133, 1992

Anderson GM, Leckman JF, Cohen DJ: Neurochemical and neuropeptide systems, in Tourette's Syndrome: Developmental Psychopathology and Clinical Care. Edited by Leckman JF, Cohen DJ. New York, Wiley, 1999, pp 261–281

Arnold PD, Siegel-Bartelt J, Cytrynbaum C, et al: Velo-cardio-facial syndrome: implications of microdeletion 22q11 for schizophrenia and mood disorders. Am J Med Genet 105:354–362, 2001

Asghari V, Sanyal S, Buchwaldt S, et al: Modulation of intracellular cyclic AMP levels by different human dopamine D4 receptor variants. J Neurochem 65:1157–1165, 1995

Auranen M, Vanhala R, Varilo T, et al: A genomewide screen for autism-spectrum disorders: evidence for a major susceptibility locus on chromosome 3q25–27. Am J Hum Genet 71:777–790, 2002

Bacanu S-A, Devlin B, Roeder K: The power of genomic control. Am J Hum Genet 66:1933–1944, 2000

Bailey A, Le Couteur A, Gottesman I, et al: Autism as a strongly genetic disorder: evidence from a British twin study. Psychol Med 25:63–78, 1995

Barr CL, Feng Y, Wigg K, et al: Identification of DNA variants in the SNAP-25 gene and linkage study of these polymorphisms and attention-deficit hyperactivity disorder. Mol Psychiatry 5:405–409, 2000

Barr C, Xu C, Kroft J, et al: Haplotype study of three polymorphisms at the dopamine transporter locus confirm linkage to attention-deficit hyperactivity disorder. Biol Psychiatry 49:333–339, 2001

Bartlett CW, Flax JF, Logue MW, et al: A major susceptibility locus for specific language impairment is located on 13q21. Am J Hum Genet 71:45–55, 2002

Battaglia M, Bertella S, Politi E, et al: Age at onset of panic disorder: influence of familial liability to the disease and of childhood separation anxiety disorder. Am J Psychiatry 152:1362–1364, 1995

Bennett AJ, Lesch KP, Heils A, et al: Early experience and serotonin transporter gene variation interact to influence primate CNS function. Mol Psychiatry 7:118–122, 2002

Biederman J, Faraone S, Keenan K, et al: Further evidence for family genetic risk factors in attention deficit hyperactivity disorder: patterns of comorbidity in probands and relatives in psychiatrically and pediatrically referred samples. Arch Gen Psychiatry 49:728–738, 1992

Birmaher B, Ryan ND, Williamson DE, et al: Childhood and adolescent depression: a review of the past 10 years, Part I. J Am Acad Child Adolesc Psychiatry 35:1427–1439, 1996

Birmaher B, Kaufman J, Brent DA, et al: Neuroendocrine response to 5-hydroxy-L-tryptophan in prepubertal children at high risk of major depressive disorder. Arch Gen Psychiatry 54:1113–1119, 1997

Birmaher B, Dahl RE, Williamson DE, et al: Growth hormone secretion in children and adolescents at high risk for major depressive disorder. Arch Gen Psychiatry 57:867–872, 2000

Blatt G, Fitzgerald C, Guptill J, et al: Density and distribution of hippocampal neurotransmitter receptors in autism: an autoradiographic study. J Autism Dev Disord 31:537–543, 2001

Boghosian-Sell L, Comings DE, Overhauser J: Tourette syndrome in a pedigree with a 7;18 translocation: identification of a YAC spanning the translocation breakpoint at 18q22.3. Am J Hum Genet 59:999–1005, 1996

Bordi F, Meller E: Enhanced behavioral stereotypies elicited by intrastriatal injection D1 and D2 dopamine agonists in intact rats. Brain Res 504:276–283, 1989

Brewerton TD, Mueller EA, Lesem MD, et al: Neuroendocrine responses to m-chlorophenylpiperazine and L-tryptophan in bulimia. Arch Gen Psychiatry 49:852–861, 1992

Bulik CM, Sullivan PF, Wade TD, et al: Twin studies of eating disorders: a review. Int J Eat Disord 27:1–20, 2000

Buxbaum JD, Silverman JM, Smith CJ, et al: Evidence for a susceptibility gene for autism on chromosome 2 and for genetic heterogeneity. Am J Hum Genet 68:1514–1520, 2001

Campbell KM, de Lecea L, Severynse DM, et al: OCD-like behaviors caused by a neuropotentiating transgene targeted to cortical and limbic D1+ neurons. J Neurosci 19:5044–5053, 1999

Cardon LR, Smith SD, Fulker DW, et al: Quantitative trait locus for reading disability on chromosome 6. Science 266:276–279, 1994

Carlson C, Papolos D, Pandita RK, et al: Molecular analysis of velocardio-facial syndrome patients with psychiatric disorders. Am J Hum Genet 60:851–859, 1997

Cases O, Seif I, Grimsby J, et al: Aggressive behavior and altered amounts of brain serotonin and norepinephrine in mice lacking MAOA. Science 268:1763–1766, 1995

Caspi A, McClay J, Moffitt TE, et al: Role of genotype in the cycle of violence in maltreated children. Science 297:851–854, 2002

Castellanos FX, Elia, J, Kruesi, MJ, et al: Cerebrospinal fluid monoamine metabolites in boys with attention-deficit hyperactivity disorder. Psychiatry Res 52:305–316, 1994

Castellanos F, Elia J, Kruesi M, et al: Cerebrospinal fluid homovanillic acid predicts behavioral response to stimulants in 45 boys with attention deficit hyperactivity disorder. Neuropsychopharmacology 14: 125–137, 1996

Cavaille J, Buiting K, Kiefmann M, et al: Identification of brain-specific and imprinted small nucleolar RNA genes exhibiting an unusual genomic organization. Proc Natl Acad Sci U S A 97:14311–14316, 2000

Chakrabarti S, Fombonne E: Pervasive developmental disorders in preschool children. JAMA 285:3093–3099, 2001

Chapman F, Visvanathan K, Carreno-Manjarrez R, et al: A flow cytometric assay for D8/17 B cell marker in patients with Tourette's syndrome and obsessive compulsive disorder. J Immunol Methods 219:181–186, 1998

Chappell P, Leckman J, Goodman W, et al: Elevated cerebrospinal fluid corticotropin-releasing factor in Tourette's syndrome: comparison to obsessive compulsive disorder and normal controls. Biol Psychiatry 39:776–783, 1996

Chen C, Rainnie DG, Greene RW, et al: Abnormal fear response and aggressive behavior in mutant mice deficient for alpha-calcium-calmodulin kinase II. Science 266:291–294, 1994

Chen L, Toth M: Fragile X mice develop sensory hyperreactivity to auditory stimuli. Neuroscience 103:1043–1050, 2001

Chen RZ, Akbarian S, Tudor M, et al: Deficiency of methyl-CpG binding protein-2 in CNS neurons results in a Rett-like phenotype in mice. Nat Genet 27:327–331, 2001

Chugani DC, Muzik O, Rothermel R, et al: Altered serotonin synthesis in the dentatothalamocortical pathway in autistic boys. Ann Neurol 42:666–669, 1997

Chugani DC, Muzik O, Behen M, et al: Developmental changes in brain serotonin synthesis capacity in autistic and nonautistic children. Ann Neurol 45:287–295, 1999

Comings DE: Blood serotonin and tryptophan in Tourette syndrome. Am J Med Genet 36:418–430, 1990

Cook EH: Autism: review of neurochemical investigation. Synapse 6: 292–308, 1990
Cook E, Leventhal B: The serotonin system in autism. Curr Opin Pediatr 8:348–354, 1996
Cook E, Arora R, Anderson G, et al: Platelet serotonin studies in hyperserotonemic relatives of children with autistic disorder. Life Sci 52: 2005–2015, 1993
Cook E, Stein M, Krasowski M, et al: Association of attention deficit disorder and the dopamine transporter gene. Am J Hum Genet 56:993–998, 1995
Cook EH Jr, Courchesne R, Lord C, et al: Evidence of linkage between the serotonin transporter and autistic disorder. Mol Psychiatry 2: 247–250, 1997a
Cook EH Jr, Lindgren V, Leventhal BL, et al: Autism or atypical autism in maternally but not paternally derived proximal 15q duplication. Am J Hum Genet 60:928–934, 1997b
Cook EH Jr., Courchesne RY, Cox NJ, et al: Linkage-disequilibrium mapping of autistic disorder, with 15q11–13 markers. Am J Hum Genet 62:1077–1083, 1998
Coplan JD, Trost RC, Owens MJ, et al: Cerebrospinal fluid concentrations of somatostatin and biogenic amines in grown primates reared by mothers exposed to manipulated foraging conditions. Arch Gen Psychiatry 55:473–477, 1998
Curran S, Mill J, Sham P, et al: QTL association analysis of the DRD4 exon 3 VNTR polymorphism in a population sample of children screened with a parent rating scale for ADHD symptoms. Am J Med Genet 105:387–393, 2001
The Cystic Fibrosis Genotype-Phenotype Consortium: Correlation between genotype and phenotype in patients with cystic fibrosis. N Engl J Med 329:1308–1313, 1993
Daly G, Hawi Z, Fitzgerald M, et al: Mapping susceptibility loci in attention deficit hyperactivity disorder: preferential transmission of parental alleles at DAT1, DBH and DRD5 to affected children. Mol Psychiatry 4:192–196, 1999
Delfs JM, Kelley AE: The role of D1 and D2 dopamine receptors in oral stereotypy induced by dopaminergic stimulation of the ventrolateral striatum. Neuroscience 39:59–67, 1990
Demitrack MA, Lesem MD, Listwak SJ, et al: CSF oxytocin in anorexia nervosa and bulimia nervosa: clinical and pathophysiologic considerations. Am J Psychiatry 147:882–886, 1990

Devlin B, Bacanu SA, Klump KL, et al: Linkage analysis of anorexia nervosa incorporating behavioral covariates. Hum Mol Genet 11:689–696, 2002

Dmitrieva TN, Oades RD, Hauffa BP, et al: Dehydroepiandrosterone sulphate and corticotropin levels are high in young male patients with conduct disorder: comparisons for growth factors, thyroid and gonadal hormones. Neuropsychobiology 43:134–140, 2001

Donnai D: Gene location in Tourette syndrome. Lancet 1:627, 1987

Dougherty DD, Bonab AA, Spencer TJ, et al: Dopamine transporter density in patients with attention deficit hyperactivity disorder. Lancet 354:2132–2133, 1999

Dresel S, Krause J, Krause KH, et al: Attention deficit hyperactivity disorder: binding of [99mTc]TRODAT-1 to the dopamine transporter before and after methylphenidate treatment. Eur J Nucl Med 27:1518–1524, 2000

Eaves L, Rutter M, Silberg JL, et al: Genetic and environmental causes of covariation in interview assessments of disruptive behavior in child and adolescent twins. Behav Genet 30:321–334, 2000

Ebstein RP, Novick O, Umansky R, et al: Dopamine D4 receptor (D4DR) exon III polymorphism associated with the human personality trait of novelty seeking. Nature Genet 12:78–80, 1996

Ernst M, Zametkin AJ, Matochik JA, et al: DOPA decarboxylase activity in attention deficit hyperactivity disorder adults. a [fluorine-18]fluorodopa PET study. J Neurosci 18:5901–5907, 1998

Ernst M, Zametkin AJ, Matochik JA, et al: High midbrain [$^{18}$F]DOPA accumulation in children with attention deficit hyperactivity disorder. Am J Psychiatry 156:1209–1215, 1999

Enard W, Przeworski M, Fisher SE, et al: Molecular evolution of FOXP2, a gene involved in speech and language. Nature 418:869–872, 2002

Faraone SV, Biederman J, Mennin D, et al: Attention-deficit hyperactivity disorder with bipolar disorder: a familial subtype? J Am Acad Child Adolesc Psychiatry 36:1378–1387, 1997

Faraone SV, Biederman J, Monuteaux MC: Toward guidelines for pedigree selection in genetic studies of attention deficit hyperactivity disorder. Genet Epidemiol 18:1–16, 2000

Faraone S, Doyle A, Mick E, et al: Meta-analysis of the association between the dopamine D4 gene 7–repeat allele and attention-deficit/hyperactivity disorder. Am J Psychiatry 158:1052–1057, 2001

Fatemi SH, Halt AR: Altered levels of Bcl2 and p53 proteins in parietal cortex reflect deranged apoptotic regulation in autism. Synapse 42: 281–284, 2001

Fatemi SH, Halt AR, Stary JM, et al: Reduction in anti-apoptotic protein Bcl-2 in autistic cerebellum. NeuroReport 12:929–933, 2001a

Fatemi SH, Stary JM, Halt AR, et al: Dysregulation of reelin and Bcl-2 proteins in autistic cerebellum. J Autism Dev Disord 31:529–535, 2001b

Fatemi S, Halt A, Stary J, et al: Glutamic acid decarboxylase 65 and 67 kDa proteins are reduced in autistic parietal and cerebellar cortices. Biol Psychiatry 52:805–810, 2002

Ferguson JN, Young LJ, Hearn EF, et al: Social amnesia in mice lacking the oxytocin gene. Nat Genet 25:284–288, 2000

Fisher SE, Marlow AJ, Lamb J, et al: A quantitative-trait locus on chromosome 6p influences different aspects of developmental dyslexia. Am J Hum Genet 64:146–156, 1999

Fisher SE, Francks C, Marlow AJ, et al: Independent genome-wide scans identify a chromosome 18 quantitative-trait locus influencing dyslexia. Nat Genet 30:86–91, 2002a

Fisher SE, Francks C, McCracken JT, et al: A genomewide scan for loci involved in attention-deficit/hyperactivity disorder. Am J Hum Genet 70:1183–1196, 2002b

Frank GK, Kaye WH, Altemus M, et al: CSF oxytocin and vasopressin levels after recovery from bulimia nervosa and anorexia nervosa, bulimic subtype. Biol Psychiatry 48:315–318, 2000

Frank GK, Kaye WH, Ladenheim EE, et al: Reduced gastrin releasing peptide in cerebrospinal fluid after recovery from bulimia nervosa. Appetite 37:9–14, 2001a

Frank GK, Kaye WH, Sahu A, et al: Could reduced cerebrospinal fluid (CSF) galanin contribute to restricted eating in anorexia nervosa? Neuropsychopharmacology 24:706–709, 2001b

Friedman L, Szabo C, Ostermeyer E, et al: Novel inherited mutations and variable expressivity of BRCA1 alleles, including the founder mutation 185delAG in Ashkenazi Jewish families. Am J Hum Genet 57:1284–1297, 1995

Frisch A, Laufer N, Danziger Y, et al: Association of anorexia nervosa with the high activity allele of the COMT gene: a family-based study in Israeli patients. Mol Psychiatry 6:243–245, 2001

Froster U, Schulte-Korne G, Hebebrand J, et al: Cosegregation of balanced translocation (1;2) with retarded speech development and dyslexia. Lancet 342:178–179, 1993

Gainetdinov RR, Wetsel WC, Jones SR, et al: Role of serotonin in the paradoxical calming effect of psychostimulants on hyperactivity. Science 283:397–401, 1999

Gayan J, Smith SD, Cherny SS, et al: Quantitative-trait locus for specific language and reading deficits on chromosome 6p. Am J Hum Genet 64:157–164, 1999

Geller B, Zimerman B, Williams M, et al: DSM-IV mania symptoms in a prepubertal and early adolescent bipolar disorder phenotype compared to attention-deficit hyperactive and normal controls. J Child Adolesc Psychopharmacol 12:11–26, 2002

George MS, Robertson MM, Costa DC, et al: Dopamine receptor availability in Tourette's syndrome. Psychiatry Res 55:193–203, 1994

Giros B, Jaber M, Jones S, et al: Hyperlocomotion and indifference to cocaine and amphetamine in mice lacking the dopamine transporter. Nature 379:606–612, 1996

Gold PW, Kaye W, Robertson GL, et al: Abnormalities in plasma and cerebrospinal-fluid arginine vasopressin in patients with anorexia nervosa. N Engl J Med 308:1117–1123, 1983

Gold PW, Gwirtsman H, Avgerinos PC, et al: Abnormal hypothalamic-pituitary-adrenal function in anorexia nervosa: pathophysiologic mechanisms in underweight and weight-corrected patients. N Engl J Med 314:1335–1342, 1986

Goodman R, Stevenson J: A twin study of hyperactivity, II: the aetiological role of genes, family relationships and perinatal adversity. J Child Psychol Psychiatry 30:691–709, 1989

Gorwood P, Ades J, Bellodi L, et al: The 5-HT(2A) -1438G/A polymorphism in anorexia nervosa: a combined analysis of 316 trios from six European centres. Mol Psychiatry 7:90–94, 2002

Green L, Fein D, Modahl C, et al: Oxytocin and autistic disorder: alterations in peptide forms. Biol Psychiatry 50:609–613, 2001

Greenough WT, Klintsova AY, Irwin SA, et al: Synaptic regulation of protein synthesis and the fragile X protein. Proc Natl Acad Sci U S A 98:7101–7106, 2001

Greer JM, Capecchi MR: Hoxb8 is required for normal grooming behavior in mice. Neuron 33:23–34, 2002

Grice DE, Leckman JF, Pauls DL, et al: Linkage disequilibrium between an allele at the dopamine D4 receptor locus and Tourette syndrome, by the transmission-disequilibrium test. Am J Hum Genet 59:644–652, 1996

Grice DE, Halmi KA, Fichter MM, et al: Evidence for a susceptibility gene for anorexia nervosa on chromosome 1. Am J Hum Genet 70: 787–792, 2002

Grigorenko EL, Wood FB, Meyer MS, et al: Susceptibility loci for distinct components of developmental dyslexia on chromosomes 6 and 15. Am J Hum Genet 60:27–39, 1997

Grigorenko EL, Wood FB, Meyer MS, et al: Linkage studies suggest a possible locus for developmental dyslexia on chromosome 1p. Am J Med Genet 105:120–129, 2001

Guy J, Hendrich B, Holmes M, et al: A mouse Mecp2-null mutation causes neurological symptoms that mimic Rett syndrome. Nat Genet 27:322–326, 2001

Hallett JJ, Harling-Berg CJ, Knopf PM, et al: Anti-striatal antibodies in Tourette syndrome cause neuronal dysfunction. J Neuroimmunol 111:195–202, 2000

Hanna GL, Yuwiler A, Cantwell DP: Whole blood serotonin in juvenile obsessive-compulsive disorder. Biol Psychiatry 29:738–744, 1991

Hanna G, Veenstra-Vander Weele J, Cox N, et al: Genome-wide linkage analysis of families with obsessive-compulsive disorder ascertained through pediatric probands. Am J Med Genet (Neuropsychiatr Genet) 114:541–552, 2002

Hebebrand J, van der Heyden J, Devos R, et al: Plasma concentrations of obese protein in anorexia nervosa. Lancet 346:1624–1625, 1995

Heinz A, Knable MB, Wolf SS, et al: Tourette's syndrome: [I-123]beta-CIT SPECT correlates of vocal tic severity. Neurology 51:1069–1074, 1998

Hess EJ, Collins KA, Wilson MC: Mouse model of hyperkinesis implicates SNAP-25 in behavioral regulation. J Neurosci 16:3104–3111, 1996

Hessl D, Dyer-Friedman J, Glaser B, et al: The influence of environmental and genetic factors on behavior problems and autistic symptoms in boys and girls with fragile X syndrome. Pediatrics 108:e88, 2001

Hoekstra PJ, Bijzet J, Limburg PC, et al: Elevated D8/17 expression on B lymphocytes, a marker of rheumatic fever, measured with flow cytometry in tic disorder patients. Am J Psychiatry 158:605–610, 2001

Holland AJ, Sicotte N, Treasure J: Anorexia nervosa: evidence for a genetic basis. J Psychosom Res 32:561–571, 1988

Holmes A, Murphy DL, Crawley JN: Reduced aggression in mice lacking the serotonin transporter. Psychopharmacology (Berl) 161:160–167, 2002

Holmes J, Payton A, Barrett J, et al: Association of DRD4 in children with ADHD and comorbid conduct problems. Am J Med Genet (Neuropsychiatr Genet) 114:150–153, 2002

Huh GS, Boulanger LM, Du H, et al: Functional requirement for class I MHC in CNS development and plasticity. Science 290:2155–2159, 2000

Hyde TM, Aaronson BA, Randolph C, et al: Relationship of birth weight to the phenotypic expression of Gilles de la Tourette's syndrome in monozygotic twins. Neurology 42:652–658, 1992

International Molecular Genetic Study of Autism Consortium: A genome wide screen for autism: strong evidence for linkage to chromosomes 2q, 7q and 16p. Am J Hum Genet 69:570–581, 2001

Jacobson, KC, Neale CA, Prescott MC, et al: Cohort differences in genetic and environmental influences on retrospective reports of conduct disorder among adult male twins. Psychol Med 30:775–787, 2000

Jimerson DC, Wolfe BE, Metzger ED, et al: Decreased serotonin function in bulimia nervosa. Arch Gen Psychiatry 54:529–534, 1997

Jones MD, Williams ME, Hess EJ: Abnormal presynaptic catecholamine regulation in a hyperactive SNAP-25-deficient mouse mutant. Pharmacol Biochem Behav 68:669–676, 2001

Jorde L, Hasstedt S, Ritvo E, et al: Complex segregation analysis of autism. Am J Hum Genet 49:932–938, 1991

Kagan J, Reznick JS, Snidman N: Biological bases of childhood shyness. Science 240:167–171, 1988

Kaplan DE, Gayan J, Ahn J, et al: Evidence for linkage and association with reading disability on 6p21.3–22. Am J Hum Genet 70:1287–1298, 2002

Karayiorgou M, Morris MA, Morrow B, et al: Schizophrenia susceptibility associated with interstitial deletions of chromosome 22q11. Proc Natl Acad Sci U S A 92:7612–7616, 1995

Karler R, Calder LD, Thai LH, et al: The dopaminergic, glutamatergic, GABAergic bases for the action of amphetamine and cocaine. Brain Res 671:100–104, 1995

Kaye WH, Berrettini W, Gwirtsman H, et al: Altered cerebrospinal fluid neuropeptide Y and peptide YY immunoreactivity in anorexia and bulimia nervosa. Arch Gen Psychiatry 47:548–556, 1990

Kaye WH, Gwirtsman HE, George DT, et al: Altered serotonin activity in anorexia nervosa after long-term weight restoration: does elevated cerebrospinal fluid 5-hydroxyindoleacetic acid level correlate with rigid and obsessive behavior? Arch Gen Psychiatry 48:556–562, 1991

Kaye W, Gendall K, Strober M: Serotonin neuronal function and selective serotonin reuptake inhibitor treatment in anorexia and bulimia nervosa. Biol Psychiatry 44:825–838, 1998

Kaye WH, Klump KL, Frank GK, et al: Anorexia and bulimia nervosa. Annu Rev Med 51:299–313, 2000

Kaye WH, Frank GK, Meltzer CC, et al: Altered serotonin 2A receptor activity in women who have recovered from bulimia nervosa. Am J Psychiatry 158:1152–1155, 2001

Kiessling LS, Marcotte AC, Culpepper L: Antineuronal antibodies in movement disorders. Pediatrics 92:39–43, 1993

Kim S-J, Cox N, Courchesne R, et al: Transmission disequilibrium mapping in the serotonin transporter gene (SLC6A4) region in autistic disorder. Mol Psychiatry 7:278–288, 2002

Koronyo-Hamaoui M, Danziger Y, Frisch A, et al: Association between anorexia nervosa and the hsKCa3 gene: a family-based and case control study. Mol Psychiatry 7:82–85, 2002

Korvatska E, Van de Water J, Anders TF, et al: Genetic and immunologic considerations in autism. Neurobiol Dis 9:107–125, 2002

Krause K, Dresel SH, Krause J, et al: Increased striatal dopamine transporter in adult patients with attention deficit hyperactivity disorder: effects of methylphenidate as measured by single photon emission computed tomography. Neurosci Lett 285:107–110, 2000

Kuikka JT, Tammela L, Karhunen L, et al: Reduced serotonin transporter binding in binge eating women. Psychopharmacology (Berl) 155: 310–314, 2001

Lahey BB, Waldman ID, McBurnett K: Annotation: the development of antisocial behavior: an integrative causal model. J Child Psychol Psychiatry 40:669–682, 1999

LaHoste G, Swanson J, Wigal S, et al: Dopamine D4 receptor gene polymorphism is associated with attention deficit hyperactivity disorder. Mol Psychiatry 1:128–131, 1996

Lai CS, Fisher SE, Hurst JA, et al: A forkhead-domain gene is mutated in a severe speech and language disorder. Nature 413:519–523, 2001

Laurino JP, Hallett J, Kiessling LS, et al: An immunoassay for antineuronal antibodies associated with involuntary repetitive movement disorders. Ann Clin Lab Sci 27:230–235, 1997

Leckman JF, Anderson GM, Cohen DJ, et al: Whole blood serotonin and tryptophan levels in Tourette's disorder: effects of acute and chronic clonidine treatment. Life Sci 35:2497–2503, 1984

Leckman JF, Goodman WK, Anderson GM, et al: Cerebrospinal fluid biogenic amines in obsessive compulsive disorder, Tourette's syndrome, and healthy controls. Neuropsychopharmacology 12:73–86, 1995

Leckman JF, Grice DE, Boardman J, et al: Symptoms of obsessive-compulsive disorder. Am J Psychiatry 154:911–917, 1997

Lee M, Martin-Ruiz C, Graham A, et al: Nicotinic receptor abnormalities in the cerebellar cortex in autism. Brain 125:1483–1495, 2002

Lenane MC, Swedo SE, Leonard H, et al: Psychiatric disorders in first degree relatives of children and adolescents with obsessive compulsive disorder. J Am Acad Child Adolesc Psychiatry 29:407–412, 1990

Leonard HL, Lenane MC, Swedo SE, et al: Tics and Tourette's disorder: a 2- to 7-year follow-up of 54 obsessive-compulsive children. Am J Psychiatry 149:1244–1251, 1992

Lesch K-P, Bengel D, Heils A, et al: Association of anxiety-related traits with a polymorphism in the serotonin transporter gene regulatory region. Science 274:1527–1531, 1996

Lijam N, Paylor R, McDonald M, et al: Social interaction and sensorimotor gating abnormalities in mice lacking Dvl1. Cell 90:895–905, 1997

Lilenfeld LR, Kaye WH, Greeno CG, et al: A controlled family study of anorexia nervosa and bulimia nervosa: psychiatric disorders in first-degree relatives and effects of proband comorbidity. Arch Gen Psychiatry 55:603–610, 1998

Lougee L, Perlmutter SJ, Nicolson R, et al: Psychiatric disorders in first-degree relatives of children with pediatric autoimmune neuropsychiatric disorders associated with streptococcal infections (PANDAS). J Am Acad Child Adolesc Psychiatry 39:1120–1126, 2000

Malison RT, McDougle CJ, van Dyck CH, et al: [123I]beta-CIT SPECT imaging of striatal dopamine transporter binding in Tourette's disorder. Am J Psychiatry 152:1359–1361, 1995

Mantzoros C, Flier JS, Lesem MD, et al: Cerebrospinal fluid leptin in anorexia nervosa: correlation with nutritional status and potential role in resistance to weight gain. J Clin Endocrinol Metab 82:1845–1851, 1997

Matsuura T, Sutcliffe JS, Fang P, et al: De novo truncating mutations in E6-AP ubiquitin-protein ligase gene (UBE3A) in Angelman syndrome. Nature Genet 15:74–77, 1997

McBride PA, Anderson GM, Hertzig ME, et al: Serotonergic responsivity in male young adults with autistic disorder. Arch Gen Psychiatry 46:205–212, 1989

McBurnett K, Lahey BB, Rathouz PJ, et al: Low salivary cortisol and persistent aggression in boys referred for disruptive behavior. Arch Gen Psychiatry 57:38–43, 2000

McCracken JT, Smalley SL, McGough JJ, et al: Evidence for linkage of a tandem duplication polymorphism upstream of the dopamine D4 receptor gene (DRD4) with attention deficit hyperactivity disorder (ADHD). Mol Psychiatry 5:531–536, 2000

McDougle C, Naylor S, Cohen D, et al: Effects of tryptophan depletion in drug-free adults with autistic disorder. Arch Gen Psychiatry 53:993–1000, 1996

McGrath MJ, Campbell KM, Parks CR, et al: Glutamatergic drugs exacerbate symptomatic behavior in a transgenic model of comorbid Tourette's syndrome and obsessive-compulsive disorder. Brain Res 877:23–30, 2000

Meaburn E, Dale PS, Craig IW, et al: Language-impaired children: no sign of the FOXP2 mutation. NeuroReport 13:1075–1077, 2002

Merette C, Brassard A, Potvin A, et al: Significant linkage for Tourette syndrome in a large French Canadian family. Am J Hum Genet 67: 1008–1013, 2000

Michelhaugh S, Fiskerstrand C, Lovejoy E, et al: The dopamine transporter gene (SLC6A3) variable number of tandem repeats domain enhances transcription in dopamine neurons. J Neurochem 79:1033–1038, 2001

Mill J, Curran S, Kent L, et al: Association study of a SNAP-25 microsatellite and attention deficit hyperactivity disorder. Am J Med Genet (Neuropsychiatr Genet) 114:269–271, 2002

Modahl C, Green L, Fein D, et al: Plasma oxytocin levels in autistic children. Biol Psychiatry 43:270–277, 1998

Moffitt T, Brammer G, Caspi A, et al: Whole blood serotonin relates to violence in an epidemiological study. Biol Psychiatry 43:446–457, 1998

Moisan MP, Courvoisier H, Bihoreau MT, et al: A major quantitative trait locus influences hyperactivity in the WKHA rat. Nat Genet 14: 471–473, 1996

Moncla A, Malzac P, Voelckel MA, et al: Phenotype-genotype correlation in 20 deletion and 20 non-deletion Angelman syndrome patients. Eur J Hum Genet 7:131–139, 1999

Morris DW, Robinson L, Turic D, et al: Family-based association mapping provides evidence for a gene for reading disability on chromosome 15q. Hum Mol Genet 9:843–848, 2000

Muller-Vahl KR, Berding G, Brucke T, et al: Dopamine transporter binding in Gilles de la Tourette syndrome. J Neurol 247:514–520, 2000a

Muller-Vahl KR, Berding G, Kolbe H, et al: Dopamine D2 receptor imaging in Gilles de la Tourette syndrome. Acta Neurol Scand 101:165–171, 2000b

Murphy KC, Jones LA, Owen MJ: High rates of schizophrenia in adults with velo-cardio-facial syndrome. Arch Gen Psychiatry 56:940–945, 1999

Murphy T, Goodman W: Genetics of childhood disorders: XXXIV. Autoimmune disorders, Part 7: D8/17 reactivity as an immunological marker of susceptibility to neuropsychiatric disorders. J Am Acad Child Adolesc Psychiatry 41:98–100, 2002

Murphy TK, Goodman WK, Fudge MW, et al: B lymphocyte antigen D8/17: a peripheral marker for childhood-onset obsessive-compulsive disorder and Tourette's syndrome? Am J Psychiatry 154:402–407, 1997

Murphy TK, Benson N, Zaytoun A, et al: Progress toward analysis of D8/17 binding to B cells in children with obsessive compulsive disorder and/or chronic tic disorder. J Neuroimmunol 120:146–151, 2001

Nelson RJ, Demas GE, Huang PL, et al: Behavioural abnormalities in male mice lacking neuronal nitric oxide synthase. Nature 378:383–386, 1995

Nelson KB, Grether JK, Croen LA, et al: Neuropeptides and neurotrophins in neonatal blood of children with autism or mental retardation. Ann Neurol 49:597–606, 2001

Nestadt G, Samuels J, Riddle M, et al: A family study of obsessive-compulsive disorder. Arch Gen Psychiatry 57:358–363, 2000

Newbury DF, Bonora E, Lamb JA, et al: FOXP2 is not a major susceptibility gene for autism or specific language impairment. Am J Hum Genet 70:1318–1327, 2002

Nicolson R, Swedo SE, Lenane M, et al: An open trial of plasma exchange in childhood-onset obsessive-compulsive disorder without poststreptococcal exacerbations. J Am Acad Child Adolesc Psychiatry 39:1313–1315, 2000

Nopola-Hemmi J, Taipale M, Haltia T, et al: Two translocations of chromosome 15q associated with dyslexia. J Med Genet 37:771–775, 2000

Nordstrom EJ, Burton FH: A transgenic model of comorbid Tourette's syndrome and obsessive-compulsive disorder circuitry. Mol Psychiatry 7:617–625, 2002

Nyhan W: The recognition of Lesch-Nyhan syndrome as an inborn error of purine metabolism. J Inherit Metab Dis 20:171–178, 1997

Nystrom-Lahti M, Parsons R, Sistonen P, et al: Mismatch repair genes on chromosomes 2p and 3p account for a major share of hereditary nonpolyposis colorectal cancer families evaluable by linkage. Am J Hum Genet 55:659–665, 1994

O'Callaghan FJ, Osborne JP: Advances in the understanding of tuberous sclerosis. Arch Dis Child 83:140–142, 2000

Orrico A, Lam C, Galli L, et al: MECP2 mutation in male patients with non-specific X-linked mental retardation. FEBS Lett 481:285–288, 2000

Pajer K, Gardner W, Rubin RT, et al: Decreased cortisol levels in adolescent girls with conduct disorder. Arch Gen Psychiatry 58:297–302, 2001

Papa M, Sellitti S, Sadile AG: Remodeling of neural networks in the anterior forebrain of an animal model of hyperactivity and attention deficits as monitored by molecular imaging probes. Neurosci Biobehav Rev 24:149–156, 2000

Pato MT, Schindler KM, Pato CN: The genetics of obsessive-compulsive disorder. Curr Psychiatry Rep 3:163–168, 2001

Pauls DL, Alsobrook JP, Goodman W, et al: A family study of obsessive-compulsive disorder. Am J Psychiatry 152:76–84, 1995

Perlmutter SJ, Leitman SF, Garvey MA, et al: Therapeutic plasma exchange and intravenous immunoglobulin for obsessive-compulsive disorder and tic disorders in childhood. Lancet 354:1153–1158, 1999

Perry EK, Lee ML, Martin-Ruiz CM, et al: Cholinergic activity in autism: abnormalities in the cerebral cortex and basal forebrain. Am J Psychiatry 158:1058–1066, 2001

Petek E, Windpassinger C, Vincent JB, et al: Disruption of a novel gene (IMMP2L) by a breakpoint in 7q31 associated with Tourette syndrome. Am J Hum Genet 68:848–858, 2001

Pine DS, Coplan JD, Wasserman GA, et al: Neuroendocrine response to fenfluramine challenge in boys: associations with aggressive behavior and adverse rearing. Arch Gen Psychiatry 54:839–846, 1997

Pine DS, Coplan JD, Papp LA, et al: Ventilatory physiology of children and adolescents with anxiety disorders. Arch Gen Psychiatry 55:123–129, 1998

Pine DS, Klein RG, Coplan JD, et al: Differential carbon dioxide sensitivity in childhood anxiety disorders and nonill comparison group. Arch Gen Psychiatry 57:960–967, 2000

Pliszka S, McCracken J, Maas J: Catecholamines in attention-deficit hyperactivity disorder: current perspectives. J Am Acad Child Adolesc Psychiatry 35:264–272, 1996

Price RA, Kidd KK, Cohen DJ, et al: A twin study of Tourette syndrome. Arch Gen Psychiatry 42:815–820, 1985

Pritchard JK, Donnelly P: Case-control studies of association in structured or admixed populations. Theor Pop Biol 60:227–237, 2001

Purcell AE, Jeon OH, Zimmerman AW, et al: Postmortem brain abnormalities of the glutamate neurotransmitter system in autism. Neurology 57:1618–1628, 2001a

Purcell AE, Rocco MM, Lenhart JA, et al: Assessment of neural cell adhesion molecule (NCAM) in autistic serum and postmortem brain. J Autism Dev Disord 31:183–194, 2001b

Rapoport JL, Fiske A: The new biology of obsessive-compulsive disorder: implications for evolutionary psychology. Perspect Biol Med 41: 159–175, 1998

Rosenberg DR, MacMaster FP, Keshavan MS, et al: Decrease in caudate glutamatergic concentrations in pediatric obsessive-compulsive disorder patients taking paroxetine. J Am Acad Child Adolesc Psychiatry 39:1096–1103, 2000

Russell V, de Villiers A, Sagvolden T, et al: Altered dopaminergic function in the prefrontal cortex, nucleus accumbens and caudate-putamen of an animal model of attention-deficit hyperactivity disorder—the spontaneously hypertensive rat. Brain Res 676:343–351, 1995

Ryan ND, Birmaher B, Perel JM, et al: Neuroendocrine response to L-5-hydroxytryptophan challenge in prepubertal major depression: depressed vs normal children. Arch Gen Psychiatry 49:843–851, 1992

Ryan ND, Dahl RE, Birmaher B, et al: Stimulatory tests of growth hormone secretion in prepubertal major depression: depressed versus normal children. J Am Acad Child Adolesc Psychiatry 33:824–833, 1994

Sagvolden T, Pettersen MB, Larsen MC: Spontaneously hypertensive rats (SHR) as a putative animal model of childhood hyperkinesis: SHR behavior compared to four other rat strains. Physiol Behav 54: 1047–1055, 1993

Sallee F, Richman H, Beach K, et al: Platelet serotonin transporter in children and adolescents with obsessive-compulsive disorder or Tourette's syndrome. J Am Acad Child Adolesc Psychiatry 35:1647–1656, 1996

Saudou F, Amara D, Dierich A, et al: Enhanced aggressive behavior in mice lacking 5-$HT_{1B}$ receptor. Science 265:1875–1878, 1994

Scerbo AS, Kolko DJ: Salivary testosterone and cortisol in disruptive children: relationship to aggressive, hyperactive, and internalizing behaviors. J Am Acad Child Adolesc Psychiatry 33:1174–1184, 1994

Schain RJ, Freedman DX: Studies on 5-hydroxyindole metabolism in autistic and other mentally retarded children. J Pediatrics 58:315–320, 1961

Schroer RJ, Phelan MC, Michaelis RC, et al: Autism and maternally derived aberrations of chromosome 15q. Am J Med Genet 76:327–336, 1998

Schulz KP, Halperin JM, Newcorn JH, et al: Plasma cortisol and aggression in boys with ADHD. J Am Acad Child Adolesc Psychiatry 36: 605–609, 1997

Schurhoff F, Bellivier F, Jouvent R, et al: Early and late onset bipolar disorders: two different forms of manic-depressive illness? J Affect Disord 58:215–221, 2000

Shahbazian M, Antalffy B, Armstrong D, et al: Insight into Rett syndrome: MeCP2 levels display tissue- and cell-specific differences and correlate with neuronal maturation. Hum Molec Genet 11:115–124, 2002a

Shahbazian M, Young J, Yuva-Paylor L, et al: Mice with truncated MeCP2 recapitulate many Rett syndrome features and display hyperacetylation of histone H3. Neuron 35:243, 2002b

Shaywitz BA, Yager RD, Klopper JH: Selective brain dopamine depletion in developing rats: an experimental model of minimal brain dysfunction. Science 191:305–308, 1976

Sherman DK, McGue MK, Iacono WG: Twin concordance for attention deficit hyperactivity disorder: a comparison of teachers' and mothers' reports. Am J Psychiatry 154:532–535, 1997

Simonic I, Gericke GS, Ott J, et al: Identification of genetic markers associated with Gilles de la Tourette syndrome in an Afrikaner population. Am J Hum Genet 63:839–846, 1998

Simonic I, Nyholt DR, Gericke GS, et al: Further evidence for linkage of Gilles de la Tourette syndrome (GTS) susceptibility loci on chromosomes 2p11, 8q22 and 11q23–24 in South African Afrikaners. Am J Med Genet 105:163–167, 2001

Simonoff E: Gene-environment interplay in oppositional defiant and conduct disorder. Child Adolesc Psychiatr Clin N Am 10:351–374, 2001

Singer HS, Hahn IH, Moran TH: Abnormal dopamine uptake sites in postmortem striatum from patients with Tourette's syndrome. Ann Neurol 30:558–562, 1991

Singer HS, Giuliano JD, Hansen BH, et al: Antibodies against human putamen in children with Tourette syndrome. Neurology 50:1618–1624, 1998

Singer HS, Giuliano JD, Hansen BH, et al: Antibodies against a neuron-like (HTB-10 neuroblastoma) cell in children with Tourette syndrome. Biol Psychiatry 46:775–780, 1999

Skuse D, James, R, Bishop, D, et al: Evidence from Turner's syndrome of an imprinted X-linked locus affecting cognitive function. Nature 387:705–708, 1997

Smalley SL, Kustanovich V, Minassian SL, et al: Genetic linkage of attention-deficit/hyperactivity disorder on chromosome 16p13, in a region implicated in autism. Am J Hum Genet 71:759–763, 2002

Smoller JW, Rosenbaum JF, Biederman J, et al: Genetic association analysis of behavioral inhibition using candidate loci from mouse models. Am J Med Genet 105:226–235, 2001

Stein DJ: Neurobiology of the obsessive-compulsive spectrum disorders. Biol Psychiatry 47:296–304, 2000

Stein MB, Uhde TW: Biology of anxiety disorders, in The American Psychiatric Press Textbook of Psychopharmacology. Edited by Schatzberg AF, Nemeroff CB. Washington, DC, American Psychiatric Press, 1995, pp 501–521

Strober M, Morrell W, Burroughs J, et al: A family study of bipolar I disorder in adolescence: early onset of symptoms linked to increased familial loading and lithium resistance. J Affect Disord 15:255–268, 1988

Strober M, Freeman R, Lampert C, et al: Controlled family study of anorexia nervosa and bulimia nervosa: evidence of shared liability and transmission of partial syndromes. Am J Psychiatry 157:393–401, 2000

Sultana R, Yu C, Yu J, et al: Identification of a novel gene on chromosome 7q11.2 interrupted by a translocation breakpoint in a pair of autistic twins. Genomics 80:129–134, 2002

Swedo SE: Sydenham's chorea: a model for childhood autoimmune neuropsychiatric disorders. JAMA 272:1788–1791, 1994

Swedo SE, Leonard HL, Mittleman BB, et al: Identification of children with pediatric autoimmune neuropsychiatric disorders associated with streptococcal infections by a marker associated with rheumatic fever. Am J Psychiatry 154:110–112, 1997

Tauscher J, Pirker W, Willeit M, et al: [123I]Beta-CIT and single photon emission computed tomography reveal reduced brain serotonin transporter availability in bulimia nervosa. Biol Psychiatry 49:326–332, 2001

Taylor JR, Morshed SA, Parveen S, et al: An animal model of Tourette's syndrome. Am J Psychiatry 159:657–660, 2002

Taylor LD, Krizman DB, Jankovic J, et al: 9p monosomy in a patient with Gilles de la Tourette's syndrome. Neurology 41:1513–1515, 1991

Terry AV Jr, Hernandez CM, Buccafusco JJ, et al: Deficits in spatial learning and nicotinic-acetylcholine receptors in older, spontaneously hypertensive rats. Neuroscience 101:357–368, 2000

Tierney E, Nwokoro NA, Porter FD, et al: Behavior phenotype in the RSH/Smith-Lemli-Opitz syndrome. Am J Med Genet 98:191–200, 2001

Todd RD, Neuman R, Geller B, et al: Genetic studies of affective disorders: should we be starting with childhood onset probands? J Am Acad Child Adolesc Psychiatry 32:1164–1171, 1993

The Tourette Syndrome Association International Consortium for Genetics: A complete genome screen in sib pairs affected by Gilles de la Tourette syndrome. Am J Hum Genet 65:1428–1436, 1999

Turjanski N, Sawle GV, Playford ED, et al: PET studies of the presynaptic and postsynaptic dopaminergic system in Tourette's syndrome. J Neurol Neurosurg Psychiatry 57:688–692, 1994

Unis A, Cook E, Vincent J, et al: Platelet serotonin measures in adolescents with conduct disorder. Biol Psychiatry 42:553–559, 1997

Urwin RE, Bennetts B, Wilcken B, et al: Anorexia nervosa (restrictive subtype) is associated with a polymorphism in the novel norepinephrine transporter gene promoter polymorphic region. Mol Psychiatry 7:652–657, 2002

U.S. Census Bureau: Resident population estimates of the United States by age and sex: April 1, 1990 to July 1, 1999, with short-term projection to November 1, 2000, 2001

Usiskin SI, Nicolson R, Krasnewich DM, et al: Velocardiofacial syndrome in childhood-onset schizophrenia. J Am Acad Child Adolesc Psychiatry 38:1536–1543, 1999

van Dyck CH, Quinlan DM, Cretella LM, et al: Unaltered dopamine transporter availability in adult attention deficit hyperactivity disorder. Am J Psychiatry 159:309–312, 2002

van Goozen SH, Matthys W, Cohen-Kettenis PT, et al: Adrenal androgens and aggression in conduct disorder prepubertal boys and normal controls. Biol Psychiatry 43:156–158, 1998

Vincent JB, Herbrick JA, Gurling HM, et al: Identification of a novel gene on chromosome 7q31 that is interrupted by a translocation breakpoint in an autistic individual. Am J Hum Genet 67:510–514, 2000

Volkow ND, Wang G, Fowler JS, et al: Therapeutic doses of oral methylphenidate significantly increase extracellular dopamine in the human brain. J Neurosci 21:RC121(1–5), 2001

Volkow ND, Wang GJ, Fowler JS, et al: Relationship between blockade of dopamine transporters by oral methylphenidate and the increases in extracellular dopamine: therapeutic implications. Synapse 43:181–187, 2002

Vukhac KL, Sankoorikal EB, Wang Y: Dopamine D2L receptor- and age-related reduction in offensive aggression. NeuroReport 12:1035–1038, 2001

Waldman I, Rowe D, Abramowitz A, et al: Association and linkage of the dopamine transporter gene and attention-deficit hyperactivity disorder in children: heterogeneity owing to diagnostic subtype and severity. Am J Hum Genet 63:1767–1776, 1998

Walsh BT, Katz JL, Levin J, et al: Adrenal activity in anorexia nervosa. Psychosom Med 40:499–506, 1978

Wan M, Lee SS, Zhang X, et al: Rett syndrome and beyond: recurrent spontaneous and familial MECP2 mutations at CpG hotspots. Am J Hum Genet 65:1520–1529, 1999

Wassink TH, Piven J, Vieland VJ, et al: Evaluation of FOXP2 as an autism susceptibility gene. Am J Med Genet 114:566–569, 2002

Wendlandt JT, Grus FH, Hansen BH, et al: Striatal antibodies in children with Tourette's syndrome: multivariate discriminant analysis of IgG repertoires. J Neuroimmunol 119:106–113, 2001

Wenk GL, Mobley SL: Choline acetyltransferase activity and vesamicol binding in Rett syndrome and in rats with nucleus basalis lesions. Neuroscience 73:79–84, 1996

Wickramaratne PJ, Warner V, Weissman MM: Selecting early onset MDD probands for genetic studies: results from a longitudinal high-risk study. Am J Med Genet 96:93–101, 2000

Willcutt EG, Pennington BF, Smith SD, et al: Quantitative trait locus for reading disability on chromosome 6p is pleiotropic for attention-deficit/hyperactivity disorder. Am J Med Genet 114:260–268, 2002

Wilson MC: Coloboma mouse mutant as an animal model of hyperkinesis and attention deficit hyperactivity disorder. Neurosci Biobehav Rev 24:51–57, 2000

Wolf SS, Jones DW, Knable MB, et al: Tourette syndrome: prediction of phenotypic variation in monozygotic twins by caudate nucleus D2 receptor binding. Science 273:1225–1227, 1996

Wong DF, Singer HS, Brandt J, et al: D2-like dopamine receptor density in Tourette syndrome measured by PET. J Nucl Med 38:1243–1247, 1997

Wozniak J, Biederman J, Mundy E, et al: A pilot family study of childhood-onset mania. J Am Acad Child Adolesc Psychiatry 34:1577–1583, 1995

Yairi E, Ambrose N, Cox N: Genetics of stuttering: a critical review. J Speech Hear Res 39:771–784, 1996

Young LJ, Nilsen R, Waymire KG, et al: Increased affiliative response to vasopressin in mice expressing the V1a receptor from a monogamous vole. Nature 400:766–768, 1999

Yu CE, Dawson G, Munson J, et al: Presence of large deletions in kindreds with autism. Am J Hum Genet 71:100–115, 2002

Zabriskie JB, Lavenchy D, Williams RC Jr, et al: Rheumatic fever-associated B cell alloantigens as identified by monoclonal antibodies. Arthritis Rheum 28:1047–1051, 1985

Zametkin AJ, Rapoport JL: Neurobiology of attention deficit disorder with hyperactivity: where have we come in 50 years? J Am Acad Child Adolesc Psychiatry 26:676–686, 1987

Zhang H, Leckman JF, Pauls DL, et al: Genomewide scan of hoarding in sib pairs in which both sibs have Gilles de la Tourette syndrome. Am J Hum Genet 70:896–904, 2002

Zhang K, Tarazi FI, Baldessarini RJ: Role of dopamine D(4) receptors in motor hyperactivity induced by neonatal 6-hydroxydopamine lesions in rats. Neuropsychopharmacology 25:624–632, 2001

Zhuang X, Oosting RS, Jones SR, et al: Hyperactivity and impaired response habituation in hyperdopaminergic mice. Proc Natl Acad Sci U S A 98:1982–1987, 2001

Chapter 2

# Molecular Genetics

## A Role in Diagnosis and Treatment of Psychiatric Disorders?

*Francis J. McMahon, M.D.*

Molecular genetic discoveries have accumulated at an unprecedented pace in the past year. The completion of the draft human genome sequence; the availability of new, high-throughput genotyping technologies; and rapid advances in functional genomics have provided human geneticists with a very powerful set of laboratory tools. For the first time, complex human diseases—those caused by a combination of genes and environmental influences—are yielding their secrets. Genetic variation conferring risk to diabetes (Horikawa et al. 2000), inflammatory bowel disease (Rioux et al. 2001), and a few other complex diseases has now been identified. The translation of these discoveries into advances in diagnosis and treatment is becoming a top research priority. Similar discoveries are beginning to appear in psychiatric disorders. How will these and future discoveries affect our diagnostic and treatment practices in the coming years?

## Genetic Basis of Major Mental Illness

Most of the major mental illnesses are known to have a significant genetic component in their etiology. This conclusion is, in most cases, supported by converging evidence from twin, family, and adoption studies. However, in no mental illness is there expected to be a one-to-one relationship between genes and disease. Instead, genes are thought of as "risk factors" that increase the probability that mental illness will occur but that do not determine it. The final outcome as to health or illness probably depends on a number of factors, including life experiences, other genes, and their interaction. This probabilistic relationship between genes and disease—the hallmark of so-called complex diseases—sets a bound on the role genetic findings will ultimately play in diagnosis and treatment. In no field is this more true than in psychiatry, in which the complex relationship between genes and disease is further complicated by the special problems we face in understanding the connection between brain diseases, on the one hand, and symptoms that manifest in the mind and behavior of the patient, on the other.

## Common Alleles for Common Diseases

The probabilistic relationship between genes and psychiatric disorders is a major factor in understanding the clinical significance of genetic findings. Another major factor is susceptibility allele frequency, which refers to the prevalence of the gene forms that contribute to the risk for illness.

How common are the alleles that contribute to risk for mental illness? To answer this question, we need to know some variables that can be measured directly, such as disease prevalence, and some that can be estimated only when all of the relevant genes have been discovered and characterized. Among the variables that we can only estimate are the number of genes involved, the degree to which each gene increases the risk for disease, and the ways in which multiple genes interact to cause disease.

It is now clear that the major mental illnesses are all very common. Large-scale epidemiological studies conducted in the United States and Europe in recent years, using modern methods of ascertainment, evaluation, and diagnosis, indicate that bipolar disorder and schizophrenia each have a lifetime prevalence of at least 1%, while phobias, substance use disorders, and major depression are even more common (Kessler et al. 1994; Robins et al. 1984).

The number of genes involved in the major mental illnesses is unclear, but data from family studies provide a basis for educated guessing. As pointed out by Risch (1990), the rate at which risk for disease falls with each step of relatedness away from the index case can provide a good estimate of the number of genes involved in a genetically complex disease. Comparisons between monozygotic and dizygotic twins can be particularly informative, since twin pairs are expected to experience more similar environments than do other types of relative pairs.

Estimates of the rates of illness among relatives come from family studies. Family study designs vary, but the best family studies attempt to directly interview all first- and second-degree relatives, perform diagnoses blind to the diagnoses of other relatives, and use epidemiological methods to adjust the observed rates of illness for differences in age among different classes of relatives. Data of this kind are currently available mainly for bipolar disorder, unipolar disorder, and schizophrenia (Baron et al. 1985; Gershon et al. 1982; Weissman et al. 1984).

For schizophrenia and bipolar disorder, the risk of illness is about 10- to 15-fold higher among the first-degree relatives of ill people than among members of the general population. However, this increased risk drops off very quickly with each step of relatedness, so that second-degree relatives have only a slightly increased risk and third-degree relatives have no measurably increased risk at all. When we compare relatives, the risk to first-degree relatives is about four times greater than the risk to second-degree relatives. If we assume that someone must have inherited all of the common susceptibility alleles to fall ill and that these alleles act together to influence disease, it is easy to show that for bipolar disorder and schizophrenia at least two but no more than five genes are likely to be involved. The existing data are less consistent in studies of unipolar disorder, but it appears that at least several genes are involved in that disorder.

The idea of several, but not many, genes interacting to cause disease, also known as the *oligogenic model,* is consistent with most other data in the field. Linkage studies, even of large samples, have failed to consistently detect single genes of major effect in any mental illness, probably because such genes do not exist or are so rare that they have eluded detection. In the case of bipolar disorder and schizophrenia, linkage studies have collectively implicated many different chromosomal regions. Individual linkage findings have not been conclusive, but several different regions have received support from more than one linkage study, suggesting that several different disease genes are indeed involved (for reviews, see Prathikanti and McMahon 2001; Sawa and Snyder 2002).

The other major variable, one on which we have the least information, is the way in which individual genes interact to influence disease. Genes can interact additively, with each allele increasing risk in a stepwise fashion. In that case, homozygotes have about twice the risk for disease as heterozygotes. Genes can also interact multiplicatively, a mechanism that is evident when the risk for disease in homozygotes is more than twice that in heterozygotes.

If we accept the idea that there are several genes that collectively contribute to the risk for bipolar disorder, with no one gene being sufficient to cause illness, we are forced to conclude that each susceptibility allele must be very common. For example, if three genes (acting in an equal and additive fashion) are needed to manifest an illness with a population prevalence of 2%, then each susceptibility allele must have a frequency equal to the cube root of 0.02, or 27% (Figure 2–1). Note that if the disease is as common as 4% and five genes (acting equally and additively) are needed to manifest the disease, then susceptibility allele frequencies can climb to more than 50%. This means that the most common form of the gene—the form seen in most people in the population—will be the form that increases risk for disease. But this increased risk will be appreciated only in that small proportion of the population that happens to inherit all five susceptibility alleles. The assumption of equal and additive effects is of course an oversimplification, but these calculations suggest that the susceptibility alleles for psychiatric disorders are more common than is widely assumed.

Of course it is possible each susceptibility allele could be individually quite rare, and that the genetic effects we observe are due to the additive influence of many different alleles. Some theoretical geneticists have argued that this scenario is not unlikely (Pritchard 2001). Such a scenario would certainly complicate gene-

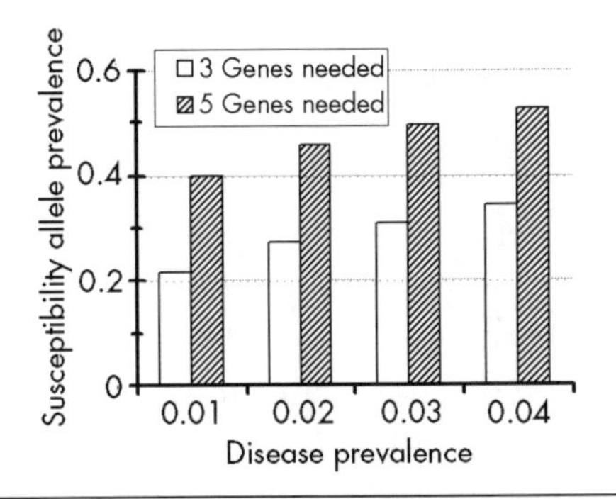


**Figure 2–1.** Theoretical susceptibility allele frequencies under a variety of additive genetic models.
The susceptibility allele frequencies are calculated as $f = p^{(1/n)}$, where $f$ is the allele frequency, $p$ is the disease prevalence, and $n$ is the number of equal and additive alleles needed to express disease.

finding efforts but would not vitiate the clinical usefulness of such genes when found. Even some of the most highly variable human genes that have been studied to date have a few alleles that are seen in the vast majority of people, regardless of ethnic background or health status.

## The Common Disease–Common Variant Hypothesis

The suggestion that susceptibility alleles for psychiatric disorders will be common is consistent with what has come to be known in the field of human genetics as the "common disease–common variant" hypothesis. Under this hypothesis, common diseases are expected to be related to common forms of genetic variation, that is, common alleles. These common alleles have typically arisen very early in human history and are widespread in all world populations. Biologically, this genetic variation is conservative, causing only subtle changes in gene function. This is in contrast to rare, Mendelian diseases that are caused by rare forms of genetic variation that tend to have arisen recently in human history, occur in particular populations, and lead to major disruptions in gene function.

A corollary to the common disease–common variant hypothesis is the idea that the higher-order structure of genetic variation in humans—variation that involves groups of related sites along a chromosome—is simple and finite (Daly et al. 2001). Genetic variation tends to occur within groups of neighboring polymorphisms, rather than at individual sites varying independently. This leads to the formation of what are called *haplotype blocks:* strings of related sites along a chromosome that tend to vary together in a population (Figure 2–2).

While many haplotypes in a given block may exist around the world, only a few haplotypes are maintained in the population at high frequencies. When viewed from this perspective, human beings become much more alike than different, and it is clear why susceptibility to common diseases is something we all share.

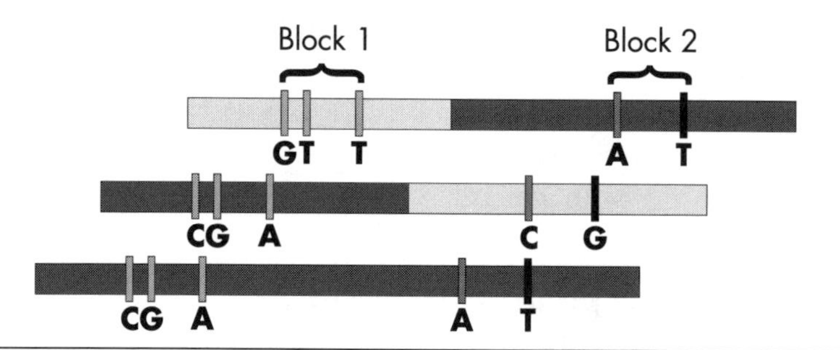


**Figure 2–2.** Haplotype blocks in the human genome.

## Translating Genetic Findings Into Clinical Practice

The progress in human genetic research toward new therapeutic discoveries can be slow. The progress of genetic research in humans is shown in Figure 2–3. The first step is the localization of the genes involved. The key milestone is reached when multiple studies point to the same chromosomal location in several different groups of patients. This is a stage we have now reached for some psychiatric disorders. The identification and positional cloning of the actual genes follow later, sometimes after laborious work with association analysis and genetic modeling in model

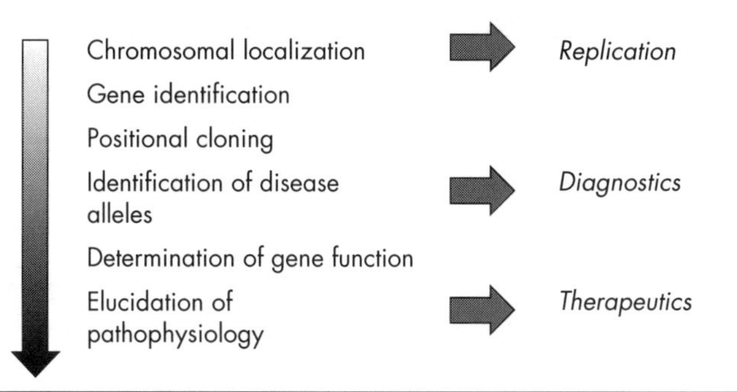


**Figure 2–3.** Progress of genetic research in humans from chromosomal localization through elucidation of pathophysiology.

organisms such as the mouse. Then comes the first milestone with immediate clinical implications: the identification of susceptibility alleles. Once this is accomplished, it is possible to develop diagnostic tests that identify carriers of susceptibility alleles. Of course, such diagnostic tests may be of little use if the alleles involved are only loosely associated with disease. This is an issue that can be addressed only when large sets of genetic data are already available.

Moving to the determination of gene function is another potentially labor-intensive step, but this step is becoming more efficient as our functional genetics knowledge continues to increase. Knowledge of gene function is only the first step in the elucidation of the pathophysiological mechanisms involved. This step, which has yet to be fully realized for any complex genetic disease, is essential if we are to begin to talk about truly gene-based therapeutics.

## Specific Applications of Genetic Findings

When genetic discoveries do come in psychiatry, they have the potential to affect our clinical practice in many ways. Among these, we discuss here six: validation of psychiatric disease entities and diagnostic classifications, improved methods of clinical diagnosis, improved treatment planning, insight into nongenetic causes, novel therapies, and the potential to develop preventive interventions.

## *Validation of Psychiatric Disease Entities and Diagnostic Classifications*

The discovery of genes will provide new standards by which to validate psychiatric disease entities. Currently, we have little information on the basis of which to validate these entities, such as bipolar disorder. In other words, we do not know whether all people with the diagnosis of bipolar disorder share some underlying biological feature(s) that differentiates them from others who do not have bipolar disorder.

Standards for validating psychiatric disease entities were proposed over 20 years ago (Robins and Guze 1970). These include clinical description, laboratory studies, natural history of illness, and familial aggregation. Unfortunately, these standards alone have not led us to disease entities that we know are valid (Hyman 2002). Gene discoveries will provide a new yardstick of validity, allowing us for the first time to compare patients and nonpatients in terms of their underlying genetic status. Such a comparison is important, since individuals who share genetic features are much more likely to share the biology in which those genes participate.

Genetic findings will also allow us for the first time to test whether our widely used disease classification systems, such as the DSM, actually correspond to underlying biological entities. The distinctions typically drawn between paranoid and disorganized forms of schizophrenia, between psychotic and nonpsychotic forms of bipolar disorder, or between early- and late-onset forms of either illness may not hold up when we understand more about the underlying genetic picture. The spinocerebellar atrophies (SCA), a group of degenerative neurological disorders, offer a good example. The discovery of most of the genes underlying these disorders has led to a complete revision of the previous diagnostic classification. Some forms of SCA previously thought to represent one disease have been split into two, each with its own genetics, while other forms of SCA previously thought to be distinct have been traced to exactly the same genetic mutation (for review, see Di Donato et al. 2001).

## *Improved Methods of Clinical Diagnosis*

Genetic discoveries may also improve diagnostic methods, allowing psychiatrists to use genetic testing to confirm clinical diagnoses, even in cases in which the presenting symptoms are subtle or complex. Does the cocaine addict who complains of depressed mood also have bipolar disorder? Is the withdrawn adolescent manifesting early signs of schizophrenia? The usefulness of genetic testing for diagnosis depends primarily on the positive predictive value of the test—that is, the proportion of all people testing positive who actually have the disease. However, since even reasonably predictive tests may vary in their potential to influence prevention or treatment efforts—not to mention in their personal and societal consequences—careful consideration will also need to be given to issues such as analytical validity (reliability, sensitivity, and specificity), clinical utility (the probability that the test will lead to a better health outcome), and the ethical, legal, and social implications of the test, whether positive or not (Burke et al. 2002).

## *Improved Treatment Planning*

Genetic discoveries will also lead to improved treatment planning. Most currently available pharmacological treatments in psychiatry are associated with average response rates of 50%–65%. Some second- and third-line therapies have even lower response rates but seem to help a distinct minority of patients. Genetic discoveries may enable the clinician to identify particular patients who are most likely to benefit from particular pharmacological interventions. On the other hand, it may also become possible to identify subgroups of patients who are more likely to experience serious adverse events, avoiding the offending agents in these patients. These kinds of genetic applications belong to the new and burgeoning fields of pharmacogenetics or pharmacogenomics (for review, see Roses 2002).

## *Insight Into Nongenetic Causes*

Genetic discoveries could also provide new insights into the fundamental causes of mental disorders; both those causes that are genetic in nature and those that are not. Genetic discoveries may allow us to study individuals who carry high-risk alleles but nevertheless do not develop psychiatric symptoms. How these people differ from those with a similar genetic endowment who do fall ill will provide important clues to the nongenetic causes of mood disorders as well as to the role of protective factors.

### *Novel Therapies*

There is also the potential for novel therapies that arise from the deeper understanding of disease physiology that genetic discoveries could provide. Genetic discoveries can provide unique insights, especially in the case of psychiatric disorders about whose underlying physiological processes we understand so little.

### *Preventive Interventions*

The study of people who are at high risk for developing mood disorders could also lead eventually to information about how to prevent mood disorders in susceptible people. Attempts at primary prevention of mental illness have up to now been unsuccessful, but this might be because these attempts were undertaken without the kind of solid information that genetic discoveries could ultimately help provide. Since a large proportion of people with mental illness, including most who die of suicide, never see a psychiatrist (Luoma et al. 2002; ten Have et al. 2002), effective primary prevention strategies could have a large impact on public mental health.

## Genetic Attributable Risk

The concept of attributable risk provides the bridge between genetic discoveries and clinical applications. In epidemiology, attributable risk has to do with the proportion of ill people who would not be ill if it were not for a particular contributing factor. Genetic attributable risk is thus the proportion of ill people who would not be ill if they did not carry a particular susceptibility allele.

We argued earlier in this chapter that several genes are likely to contribute to each major psychiatric disorder and that each susceptibility allele is expected to be very common, even though each allele may individually increase risk for illness only slightly. The concept of genetic attributable risk shows that even alleles that cause only a small increase in an individual's disease risk can have a large impact when those alleles are frequent in the population. Thus, the clinical impact of genetic discoveries in psychiatric research is expected to be large, even if the clinical impact of the individual alleles is not.

## Conclusion: The Psychiatrist as Interpreter of Genetic Findings

The accelerated pace of genetic discoveries in psychiatry will increasingly challenge clinical psychiatrists. Not only do we need to stay abreast of research developments, but we increasingly need to interpret these findings for our patients and their families. This is an essential role, since the psychiatrist is uniquely placed to put research findings into an appropriate medical and clinical context that will maximize the potential benefit to patients while minimizing the risks that arise from misconceptions and misunderstanding of this complex subject matter. Genetically based diagnostic testing, treatment planning, and preventive medicine will pose even greater challenges that will again call on psychiatrists to play a unique role, translating genetic advances into effective and humane medical care.

## References

Baron M, Gruen R, Rainer JD, et al: A family study of schizophrenic and normal control probands: implications for the spectrum concept of schizophrenia. Am J Psychiatry 142:447–455, 1985

Burke W, Atkins D, Gwinn M, et al: Genetic test evaluation: information needs of clinicians, policy makers, and the public. Am J Epidemiol 156:311–318, 2002

Daly MJ, Rioux JD, Schaffner SF, et al: High-resolution haplotype structure in the human genome. Nat Genet 29:229–232, 2001

Di Donato S, Gellera C, Mariotti C: The complex clinical and genetic classification of inherited ataxias, II: autosomal recessive ataxias. Neurol Sci 22:219–228, 2001

Gershon ES, Hamovit J, Guroff JJ, et al: A family study of schizoaffective, bipolar I, bipolar II, unipolar, and normal control probands. Arch Gen Psychiatry 39:1157–1167, 1982

Horikawa Y, Oda N, Cox NJ, et al: Genetic variation in the gene encoding calpain-10 is associated with type 2 diabetes mellitus. Nat Genet 26:163–175, 2000

Hyman SE: Neuroscience, genetics, and the future of psychiatric diagnosis. Psychopathology 35:139–144, 2002

Kessler RC, McGonagle KA, Zhao S, et al: Lifetime and 12-month prevalence of DSM-III-R psychiatric disorders in the United States: results from the National Comorbidity Survey. Arch Gen Psychiatry 51:8–19, 1994

Luoma JB, Martin CE, Pearson JL: Contact with mental health and primary care providers before suicide: a review of the evidence. Am J Psychiatry 159:909–916, 2002

Prathikanti S, McMahon FJ: Genome scans for susceptibility genes in bipolar affective disorder. Ann Med 33:257–262, 2001

Pritchard JK: Are rare variants responsible for susceptibility to complex diseases? Am J Hum Genet 69:124–137, 2001

Rioux JD, Daly MJ, Silverberg MS, et al: Genetic variation in the 5q31 cytokine gene cluster confers susceptibility to Crohn disease. Nat Genet 29:223–228, 2001

Risch N: Linkage strategies for genetically complex traits. Am J Hum Genet 46:229–241, 1990

Robins E, Guze SB: Establishment of diagnostic validity in psychiatric illness: its application to schizophrenia. Am J Psychiatry 126:983–987, 1970

Robins LN, Helzer JE, Weissman MM, et al: Lifetime prevalence of specific psychiatric disorders in three sites. Arch Gen Psychiatry 41:949–958, 1984

Roses AD: Pharmacogenetics place in modern medical science and practice. Life Sci 70:1471–1480, 2002

Sawa A, Snyder SH: Schizophrenia: diverse approaches to a complex disease. Science 296:692–695, 2002

ten Have M, Vollebergh W, Bijl R, et al: Bipolar disorder in the general population in The Netherlands (prevalence, consequences and care utilisation): results from The Netherlands Mental Health Survey and Incidence Study (NEMESIS). J Affect Disord 68(2–3):203–213, 2002

Weissman MM, Gershon ES, Kidd KK, et al: Psychiatric disorders in the relatives of probands with affective disorders. Arch Gen Psychiatry 41:13–21, 1984

Chapter 3

# Molecular Neurobiology and Schizophrenia

## Implications for Etiology and Treatment

*Andrew R. Gilbert, M.D.*
*David W. Volk, Ph.D.*
*David A. Lewis, M.D.*

Schizophrenia is a complex disorder with a pathogenesis and pathophysiology that remain to be elucidated. With a 1% lifetime incidence and substantial morbidity, schizophrenia is a widespread and disabling condition. As the fields of neuroscience and behavioral science evolve, the diagnosis, treatment, and understanding of schizophrenia continue to develop. The heterogeneity characteristic of the clinical presentation of schizophrenia is consistent with the variety of neurobiological abnormalities observed in affected individuals, which appear to arise from the interactions of multiple genetic and environmental factors. In this chapter, we summarize recent findings from research on the neurobiology of schizophrenia. First, we provide a brief overview of the genetic, environmental, and developmental factors that appear to be involved in the pathogenesis of schizophrenia. Second, we present four examples of susceptibility genes that may contribute to the liability for schizophrenia. Third, we consider the alterations in several molecular systems that appear to be relevant to the pathophysiology of schizophrenia. Finally, we discuss the implications of these findings which, when synthesized, may both inform etiopathogenic models of the illness and suggest novel approaches to its treatment.

## Etiological Factors and Schizophrenia

### *Genetic Risk Factors*

Although the roles of specific susceptibility genes remain elusive, inheritance clearly contributes to the etiology of schizophrenia. Family, twin, and adoption studies indicate that the morbid risk of schizophrenia in the relatives of affected individuals increases with the percentage of shared genes (see Gottesman 1991; Lewis and Levitt 2002 for review). For example, although the incidence of schizophrenia in the general population is 1%, schizophrenia occurs in approximately 2% of the third-degree relatives (e.g., first cousins), in 2%–6% of the second-degree relatives (e.g., nieces/nephews), and in 6%–17% of the first-degree relatives (e.g., parents, siblings, or children) of an individual with schizophrenia. Furthermore, if one twin has the illness, the risk of schizophrenia in the other twin is approximately 17% for fraternal twins and approaches 50% for identical twins. This familial feature of schizophrenia does not appear to be due to shared environmental factors, as indicated by the observation that when the biological children of individuals with schizophrenia are adopted away, their risk of developing schizophrenia remains much higher than the general population rates of schizophrenia present in their adoptive families (Lewis and Lieberman 2000). Furthermore, for identical twins discordant for schizophrenia, their offspring have elevated rates of the disorder, regardless of whether the parent was affected or unaffected.

Although the etiology of schizophrenia clearly involves genetic factors, about 60% of all persons with schizophrenia have neither a first- nor a second-degree relative with the disorder (Gottesman 1991). Furthermore, given that the concordance for schizophrenia in monozygotic twins is only about 50% (Gottesman 1991), genetic liability alone cannot account for the clinical manifestations of the disorder. Thus, it appears that the genetic predisposition to schizophrenia is complex and that additional, nongenetic factors are required for the inherited risk to become evident.

## *Environmental Risk Factors*

A variety of environmental events, especially during early development, have been associated with an increased risk of schizophrenia. Individuals with schizophrenia have been found to have a greater frequency of adverse in utero events, such as severe maternal malnutrition during the first trimester or maternal influenza or rubella during midpregnancy (Brown and Susser 2002; Lewis and Levitt 2002). In addition, complications during labor or delivery may increase the risk for schizophrenia. For example, a recent meta-analysis found significant differences between schizophrenic subjects and comparison subjects in the following variables (in descending order of effect size): diabetes during pregnancy, birth weight < 2000 g, emergency cesarean section, congenital malformations, uterine atony, rhesus variables, asphyxia, bleeding in pregnancy, and preeclampsia (Cannon et al. 2002). A common underlying mechanism associated with these variables may involve fetal hypoxia. Interestingly, across probands with schizophrenia or schizoaffective disorder, their nonpsychotic full siblings, and a demographically similar group of healthy individuals without a family history of psychosis, hippocampal volumes were smallest in the affected individuals, intermediate in their unaffected full siblings, and largest in the healthy comparison subjects, consistent with an inverse relationship between hippocampal volume and genetic load for schizophrenia (van Erp et al. 2002). In addition, within the schizophrenia group, hippocampal volumes were smaller in those individuals who had experienced fetal hypoxia, whereas no such effect was found in the other two subject groups. Because the frequency of hypoxia-related insults was the same across all three subject groups, the findings may be most consistent with an interactive, "two-hit" model of genetic predisposition and fetal hypoxia.

The notion of hypoxic-ischemic damage as a causal risk factor for schizophrenia is limited, however, by the inability to establish independence of individual obstetric complications. In other words, the condition may be secondary to pregnancy complications, preexisting problems with the fetus, or maternal behaviors (Cannon et al. 2002). A Finnish cohort study reported that low birth weight and/or a short gestation were more common among schizophrenic subjects, suggesting that abnormal fetal development, as opposed to one independent risk factor, may be the mechanism of an obstetric role in schizophrenia risk (Jones et al. 1998). Because of small effect sizes for the relationship between obstetric complications and later schizophrenia, future studies with larger cohorts and more modern paradigms and statistical techniques may be necessary to better elucidate this environmental association (Cannon et al. 2002).

In addition to obstetric complications, factors associated with place of birth and rearing may contribute to the risk of schizophrenia (Pedersen and Mortensen 2001). For example, urban births are associated with approximately twice the risk of schizophrenia as rural births and may account for a much greater percentage of the affected population than genetic liability. Specific urban environmental factors that may influence schizophrenia risk include greater exposure to infection, toxins, or malnutrition. However, it is important to note that most individuals with these life events do not develop schizophrenia.

Another potential risk factor for adult schizophrenia may be advanced paternal age. Using a large Israeli population-based birth cohort, Malaspina et al. (2001) found a robust and "dose-related" effect of paternal age on risk of schizophrenia. Brown et al. (2002), employing more rigorous research design techniques to reduce exposure misclassification and selection bias, reported a similarly significant association. The authors of these studies posited that the risk for schizophrenia may rise because of accumulating mutations in paternal germ cells. As a result of increasing spermatocyte divisions, de novo genetic mutations due to replication errors and defective DNA repair mechanisms may develop with advancing paternal age.

## *Developmental Processes*

The temporal delay between environmental events of possible etiological significance and the appearance of clinical illness has played a major role in the idea that schizophrenia may be a disorder of neural development (Lewis and Levitt 2002). In addition, the numerous observations of subtle disturbances in motor, cognitive, and social functions during childhood and adolescence of individuals who later manifest schizophrenia have been viewed as evidence of disturbances in neurodevelopmental processes.

During the past two decades, two primary views of a neurodevelopmental model of schizophrenia have emerged, suggesting that the illness results from either early (pre- or perinatal) brain insults or late (adolescent) brain dysmaturation. The *early* models posit that a static lesion present from the pre- or perinatal period interacts with normal brain development, leading to a long latency until the appearance of clinical signs and symptoms (Weinberger 1987). Findings that have been advanced in support of an early lesion are as follows: 1) postmortem findings of aberrant cortical cytoarchitecture, suggestive of possible abnormal neuronal migration; 2) a childhood history of neurological abnormalities in preschizophrenic subjects, suggesting an association with neurointegrative deficits; 3) associations between early environmental events and illness development (reviewed in previous section); and 4) structural and functional neuroimaging findings in the high-risk offspring of schizophrenic patients, suggesting brain abnormalities prior to the onset of diagnostic features of the illness (Keshavan and Hogarty 1999).

Other findings have challenged the model of an early onset neurodevelopmental insult as a cause of schizophrenia. For example, postmortem studies have failed to confirm the cytoarchitectural disturbances in the entorhinal cortex of schizophrenic subjects, originally suggestive of aberrant neuronal migration (Harrison and Lewis, in press). Furthermore, the presence of excessive extraventricular cerebrospinal fluid (CSF), a structural feature of schizophrenia, is difficult to explain by an early, static lesion model (Woods 1999). Diffuse loss of brain tissue in the pre- or perinatal periods would produce a smaller cranial cavity and a persistent increase in ventricular size, but not an increase in extracerebral CSF. Therefore, the findings of smaller head size at birth and larger extracerebral CSF volumes in schizophrenia would likely be secondary to both an early lesion and later volume loss (Woods 1999).

Feinberg (1982), in contrast, proposed a *late* brain lesion model of limited duration and short latency, characterized by an abnormality in periadolescent pruning of cortical synapses. Abnormal brain structural volumes reported in several imaging studies of first-episode, neuroleptic-naive patients with schizophrenia may reflect aberrant synaptic pruning mechanisms (see Shenton et al. 2001 for review). Recent postmortem findings of deficits in synaptic components, suggesting reduced cortical synaptic density, may further support this model (Lewis 2002; Selemon and Goldman-Rakic 1999).

Could both early and late developmental events contribute to the pathogenesis of schizophrenia? Certainly, the available data do not exclude the possibility of pathogenic events that occur throughout development and, perhaps, into the early stages of the illness (Marenco and Weinberger 2000). As previously noted, the combination of smaller head size at birth and larger extracerebral CSF volumes may suggest both early and later developmental deficits in schizophrenia. In addition, recent evidence of accelerated gray matter loss in patients with early-onset schizophrenia, as well as suggestions of progressive brain volume changes in adult patients with schizophrenia (Lewis and Lieberman 2000), may reflect a continuing pathogenesis.

The view of schizophrenia as a developmental disorder has also been inferred from the failure to find evidence of gliosis, whether examined by Nissl staining or immunoreactivity, in the postmortem brains of individuals with schizophrenia (Harrison and Lewis, in press). This apparent absence of gliosis has been interpreted to exclude typical neurodegenerative processes as operative in schizophrenia. Given the epidemiological data for perinatal brain damage as a precursor of the illness, the absence of gliosis seems surprising from a neurodevelopmental perspective, given that the brain may be able to mount a gliotic response as early as the second trimester of gestation. On the other hand, loss of neurons and other types of disturbances may transpire in either the developing or adult brain without a sustained glial reaction (Lewis and Levitt 2002). Consequently, the absence of gliosis may be informative only regarding the possible mechanisms of brain abnormalities in schizophrenia, and not their timing.

### *Summary*

Thus, the etiopathogenesis of schizophrenia appears to involve the interplay of polygenetic influences and environmental risk factors operating on brain maturational processes (Lewis and Levitt 2002). The complexity of these potential interactions clearly complicates the study of the neurobiology of the illness (Lewis and Lieberman 2000). Schizophrenia may be a single disease process with a range of severity and clinical manifestations across individuals, determined by the extent to which different brain regions or circuits are affected. Alternatively, the clinical heterogeneity may indicate that what we recognize as schizophrenia represents, in fact, a constellation of different disease processes that share some phenotypic features. Thus, a comprehensive understanding of the molecular neurobiology of schizophrenia continues to be a future goal. Consequently, the following sections highlight interesting, but still provisional, findings on candidate genetic liabilities and molecular system alterations in schizophrenia.

## Genetic Variation and Schizophrenia

The genetic liability to schizophrenia appears to be transmitted in a polygenic, non-Mendelian fashion (see Volk and Lewis, in press for review). A number of chromosomal loci have been found to be associated with schizophrenia, including 22q11–13, 6p, 13q, and 1q21–22. Interestingly, several of these loci contain genes that have well-delineated neurobiological functions, including genes that regulate some of the molecular systems discussed in the following subsections. Although some of these loci are now associated with single-gene findings, these observations await independent replications. The following four examples illustrate different strategies for identifying candidate genes and the roles they may play in the disease process of schizophrenia.

### *COMT*

Allelic variations in genes whose protein products may contribute to the regulation of brain processes that are deficient in schizophrenia, such as cognitive functioning, may reveal susceptibility genes for this disorder. As discussed in more detail below, dopamine (DA) neurotransmission in the prefrontal cortex (PFC) plays an essential role in mediating the types of working memory cognitive processes that are disrupted in schizophrenia. Interestingly, several lines of evidence suggest that allelic variations in catechol-*O*-methyltransferase (COMT), a major enzyme involved in the metabolic degradation of PFC DA, may contribute to the liability to schizophrenia. For example, microdeletions at the locus of the COMT gene, chromosome 22q11, have been reported in approximately 2% of subjects with schizophrenia (Karayiorgou et al. 1995). Furthermore, an increased prevalence of psychotic symptoms and schizophrenia has been reported in patients with velocardiofacial syndrome, which involves variable deletions of 22q11 (Karayiorgou et al. 1995). In addition, linkage studies of loci at 22q12–13, near the locus for COMT, have also found associations with schizophrenia (for a review, see Riley and McGuffin 2000).

Polymorphisms in the COMT gene result in profound differences in the ability of COMT to metabolize DA. A single guanine to adenosine transition results in a change in amino acids from valine (val) to methionine (met) at a specific site. The met-containing COMT enzyme has 25% of the activity of val-containing COMT, leading to a marked reduction in DA metabolism for the met-containing COMT enzyme and, consequently, a prolonged presence of DA. Weinberger and colleagues (2001) have hypothesized that the presence of the high-activity val-containing COMT allele, which results in lower DA levels, may contribute to impaired cognitive function in schizophrenia. Indeed, COMT appears to play a primary role in DA metabolism in the PFC but not in other brain regions such as the striatum. For example, in COMT knockout mice, DA levels are elevated by over 200% in frontal cortex but not in striatum (Gogos et al. 1998). Furthermore, individuals with schizophrenia who are homozygous for the val genotype commit more perseverative errors on the Wisconsin Card Sort Task, which assesses working memory and other executive functions, than either heterozygotes or individuals homozygous for the met genotype (Egan et al. 2001), consistent with other studies demonstrating that a reduction in DA in the PFC impairs working memory function (Goldman-Rakic 1994). Furthermore, several family-based association studies utilizing the transmission disequilibrium test have reported weak, yet significant, associations between the val-containing COMT allele and schizophrenia (Egan et al. 2001).

However, other lines of evidence suggest that allelic variation of the COMT gene may play a relatively minor role in the pathophysiology of schizophrenia. For example, numerous case-control association studies have consistently failed to find an association between the val-containing COMT allele and schizophrenia (see Chen et al. 1997; Daniels et al. 1996; Karayiorgou et al. 1998; Liou et al. 2001). Furthermore, studies suggesting that deletion of the COMT gene, such as in velocardiofacial syndrome, may be associated with schizophrenia seem to conflict with other studies reporting an association between the high-activity val-containing COMT allele and schizophrenia. Finally, the high prevalence of the val-containing COMT allele in the general population suggests that, although the val-containing COMT allele may contribute to the presence of cognitive abnormalities in schizophrenia, its effect is notable only in the presence of other predisposing factors.

### $\alpha_7$ *Nicotinic Receptor*

Another approach to identifying susceptibility genes has exploited the presence of sensory gating abnormalities in individuals with schizophrenia (see Braff and Geyer 1989 for review). That is, affected individuals appear to have deficits in the ability to gate, or internally screen, sensory stimuli, such that only the relevant stimuli are attended to. In humans, the gating of auditory stimuli can be measured as a decrement in the evoked electroencephalographic response to the second of two consecutive auditory stimuli. Specifically, when two paired tones are delivered 500 msec apart, the amplitude of the P50 auditory evoked potential (a positive response that occurs 50 msec after a tone) to the second tone normally decreases to less than 40% of the P50 following the first tone. In contrast, in about three-quarters of individuals with schizophrenia, and in about one-half of their unaffected family members, the response to the second tone does not exhibit the expected decrease in amplitude and may actually increase (Weiland et al. 2000).

This deficit in the P50 auditory evoked potential is transmitted as an autosomal dominant in families with schizophrenia, suggesting that the power of genetic analyses for this phenotype would be increased relative to schizophrenia itself. Indeed, Freedman and colleagues (1997) found that this trait was linked to a polymorphism on chromosome 15q14. Evidence for linkage of the locus to schizophrenia was also positive, but not as strong, and, as expected, attempts to replicate the linkage to schizophrenia have produced both positive and negative findings.

The relevance of this locus for the pathophysiological mechanism of auditory gating deficits, and possibly for the pathogenesis of schizophrenia, is supported by the fact that the gene *(CHRNA7)* for the $\alpha_7$ subunit of the nicotinic acetylcholine receptor (nAChR) is located near the chromosome 15 marker (Freedman et al. 1997; Weiland et al. 2000). Several lines of evidence, in both humans and animal models, support a role for the nAChR $\alpha_7$ subunit in the pathophysiology of sensory gating deficits in schizophrenia. In humans, the binding of α-bungarotoxin to nAChRs is reduced in the postmortem hippocampus and other brain regions of individuals with schizophrenia (Freedman et al. 2001). In addition, the P50 deficit is improved by nicotine in both subjects with schizophrenia and their unaffected family members (Weiland et al. 2000). However, in contrast to control subjects, who show an upregulation of nicotinic receptors in association with smoking, individuals with schizophrenia exhibit lower binding at every level of smoking history (Breese et al. 2000). Furthermore, in rats, both the administration of specific antagonists of the $\alpha_7$ nAChR and the use of antisense oligonucleotides complementary to the $\alpha_7$ translation start site blocked sensory gating (Leonard et al. 1996).

These observations are of particular clinical interest because smoking and other uses of nicotine-containing tobacco products are much more common in individuals with schizophrenia than in the general population, and individuals with schizophrenia appear to extract more nicotine from each cigarette than do unaffected smokers (Olincy et al. 1997). These observations suggest that individuals with schizophrenia use tobacco excessively as a means of self-medication. That is, by stimulating deficient $\alpha_7$-containing nAChRs, they are able to transiently reduce the subjective distress associated with sensory gating disturbances (Weiland et al. 2000). However, how the apparent inherited liability at chromosome 15q14 is translated into the functional deficit remains unclear, and both the underexpression and an early developmental overexpression of the *CHRNA7* gene have been hypothesized (Weiland et al. 2000).

## *RGS4*

Genes that contribute to the inherited susceptibility for schizophrenia are likely to have several properties: 1) a protein product that plays a role in the pathophysiology of the illness and in the pathways that are affected by therapeutic interventions; 2) a mRNA that shows altered expression levels across the brain in the disease state; and 3) a chromosomal locus implicated in linkage and association studies. The gene for RGS4, a member of the family of regulators of G-protein signaling proteins (Berman et al. 1996), appears to be such a candidate. First, like other RGS proteins, RGS4 plays a critical role in signaling through G-protein coupled neurotransmitter receptors (GPCRs) by regulating the duration of G-protein–mediated intracellular signaling (De Vries et al. 2000). Many neurotransmitter systems that have been implicated in the pathophysiology of schizophrenia, such as $\gamma$-aminobutyric acid (GABA), glutamate, and DA (see below), utilize GPCRs that are regulated by RGS proteins, and at least some of these GPCRs are the targets of antipsychotic drugs.

Second, a recent microarray study revealed a robust and consistent decrease in cortical levels of the *RGS4* transcript in schizophrenia, whereas other RGS or G-protein signaling components represented on the microarrays did not exhibit significant transcript changes (Mirnics et al. 2001b). Furthermore, *RGS4* transcript levels were not altered in subjects with major depressive disorder or in monkeys treated chronically with haloperidol (Mirnics et al. 2001b), and in subjects with alcoholism mRNA levels of RGS4 were actually increased. Together, these findings suggest that decreased *RGS4* mRNA levels may not be a common feature of all psychiatric disorders or the result of treatment or comorbid conditions that frequently accompany schizophrenia.

Third, the *RGS4* gene is located on chromosome 1q21–22, a locus recently implicated in schizophrenia (Brzustowicz et al. 2000). The idea that variants in *RGS4* may contribute to these inheritance patterns is supported, to some degree, by the finding that of the 70 genes mapped to the 1q21–22 locus represented on the platform used in the microarray study, only *RGS4* exhibited altered expression across multiple schizophrenic subjects (Mirnics et al. 2001b). However, sequence-based analysis of the coding region of *RGS4* in these schizophrenic subjects did not report informative polymorphisms or mutations.

To determine whether altered *RGS4* expression reflects a primary inherited abnormality in schizophrenia, Chowdari and colleagues (2002) conducted genetic association studies in more than 1,400 subjects using RGS4 polymorphisms. When a genomic region of approximately 300 kb was surveyed, associations were detected for four single nucleotide polymorphisms (SNPs) localized to a 10-kb span at *RGS4*. The global association tests for haplotypes bearing all four of these SNPs suggested that the transmission distortion was unlikely to be due to chance alone, especially when all family-based samples were considered. An independent replication of this finding has also been reported. However, the specificity of the finding for schizophrenia requires further study.

## *Dysbindin*

Genetic analyses of the Irish Study of High-Density Schizophrenia Families, which includes 270 pedigrees, revealed significant associations between variations in the gene *DTNBP1* (the dystrobrevin-binding protein 1, or dysbindin, gene) and schizophrenia and related phenotypes (Straub et al. 2002). The gene for DTNBP1 is located at 6p24–21, a susceptibility locus for schizophrenia observed in previous studies.

Dysbindin is likely a protein component of the dystrophin complex (DPC) found in postsynaptic densities. The DPC plays an important role in neuromuscular synapse formation and maintenance. Dysbindin also appears to affect synaptic activity in the brain and to modulate receptors, such as $GABA_A$ receptor subtypes in the hippocampus, cortex, and cerebellum. Dysbindin's potential role in synaptic transmission and its association with schizophrenia make it a strong candidate as a susceptibility locus that may influence the risk of schizophrenia. Recent evidence of synaptic dysfunction in schizophrenia (see below) points to the importance of exploring etiologic factors, such as genes that code for synaptic proteins. Furthermore, the conceivable role of dysbindin in GABA transmission, as discussed in the next section, may further implicate its involvement in the neurobiology of schizophrenia.

## Molecular Systems and Schizophrenia: Molecular Alterations in the Context of Neural Circuitry

To understand how genetically or environmentally induced alterations in molecular systems produce disturbed brain function in schizophrenia, the actions of the molecules of interest must be considered within the constraints provided by the neural circuits that produce or utilize them (Lewis 2002). Clues regarding the affected neural circuits come from reports of structural brain abnormalities in individuals with schizophrenia (Harrison and Lewis, in press), such as the reduced volume of the cortical gray matter. These cortical changes do not represent a uniform abnormality but are found predominantly in certain association cortices, including those located in the dorsal prefrontal, the superior temporal gyrus, and limbic areas such as the hippocampal formation and anterior cingulate cortex (Lewis and Lieberman 2000; McCarley et al. 1999). These structural abnormalities have been observed in first-episode, never-medicated subjects with schizophrenia and may be present prior to the clinical onset of illness, suggesting that they reflect the primary disease process and are not a secondary consequence of the illness or of its treatment.

Interestingly, subjects with schizophrenia perform poorly on certain cognitive tasks, such as those requiring working memory (i.e., the ability to keep in mind transiently a bit of information in order to guide behavior or thought), that are subserved by circuitry involving the dorsal PFC (Goldman-Rakic 1994). In addition, these subjects fail to show normal activation of this brain region when attempting to perform such tasks (Weinberger et al. 2001). Disturbances in working memory and related cognitive functions may be the most persistent and, over the long term, most debilitating symptoms of schizophrenia. In addition, the long-term prognosis for individuals with schizophrenia appears to be best predicted by the degree of cognitive impairment (Green 1996). Thus, understanding the molecular basis for abnormalities in PFC circuitry may be particularly important for improving clinical outcome.

Postmortem studies have reported a 5%–10% reduction in cortical thickness, with a corresponding increase in cell packing density but no change in total neuron number, in the dorsal PFC of subjects with schizophrenia (Harrison and Lewis, in press; Selemon and Goldman-Rakic 1999). Although the size of some PFC neuronal populations, particularly pyramidal neurons in deep layer 3, is smaller in schizophrenic subjects (Pierri et al. 2001; Rajkowska et al. 1998), the increase in cell packing density is also likely to reflect a decrease in the number of axon terminals, distal dendrites, and dendritic spines that represent the principal components of cortical synapses. Indeed, the density of basilar dendritic spines on PFC layer 3 pyramidal neurons has been reported to be decreased in subjects with schizophrenia (Harrison and Lewis, in press). Furthermore, levels of synaptophysin, a presynaptic terminal protein, are decreased in the PFC of subjects with schizophrenia in multiple studies (Lewis and Lieberman 2000). Finally, levels of *N*-acetylaspartate, a marker of axonal and/or neuronal integrity, are reduced in the PFC of subjects with schizophrenia (Weinberger et al. 2001). Thus, convergent lines of evidence from neuroimaging and postmortem studies suggest that schizophrenia is associated with synaptic alterations in the PFC.

Consequently, in the following subsections, we review findings that address the molecular bases and consequences of synaptic-related abnormalities in the PFC in schizophrenia.

## *Alterations in the Machinery and Metabolic Support of PFC Synapses*

Gene chips or microarrays represent a powerful approach to determine, simultaneously, the level of expression of thousands of genes by measuring the tissue concentrations of their mRNAs. When this technology was used, in the postmortem dorsal PFC of subjects with schizophrenia the most consistently altered group of genes (of more than 250 gene groups examined) included those whose protein products are involved in the machinery of presynaptic function (Mirnics et al. 2000). That is, decreased tissue mRNA levels were found for proteins located on or associated with synaptic vesicles or the presynaptic membrane, and that are known or hypothesized to play a role in the release of neurotransmitters. Although the extent to which the reported transcript changes are converted into alterations at the protein level remains to be determined, other studies of individual synapse-associated proteins (e.g., synaptophysin, SNAP-25) have generally found reductions in the PFC of subjects with schizophrenia.

The disturbances in dorsal PFC synaptic function in schizophrenia also appear to be associated with altered metabolism in this brain region. For example, the normal increases in glucose utilization and blood flow present in the dorsal PFC when subjects are asked to perform certain cognitive tasks are markedly blunted in subjects with schizophrenia, compared with the large activations and blood flow increases seen in normal subjects (Weinberger et al. 2001). Using a microarray approach, Middleton and colleagues (2002) examined the expression profile of genes that make up major metabolic pathways in the dorsal PFC of subjects with schizophrenia. Of the 71 metabolic pathways assessed, only five showed consistent changes in expression in the subjects with schizophrenia. Specifically, reduced expression levels were identified for the transcripts of genes whose protein products are involved in the regulation of ornithine and polyamine metabolism, the mitochondrial malate shuttle system, the transcarboxylic acid (TCA) cycle, aspartate and alanine metabolism, and ubiquitin metabolism. Parallel in situ hybridization studies were also conducted in macaque monkeys treated chronically with haloperidol in a manner that mimicked the clinical use of this antipsychotic medication. Interestingly, although most of the metabolic genes that showed decreased expression in the subjects with schizophrenia were not similarly altered in the monkeys, the transcript encoding the cytosolic form of malate dehydrogenase displayed a marked haloperidol-associated *increase* in expression. These findings suggest that metabolic alterations in the PFC of subjects with schizophrenia reflect highly specific abnormalities, at least at the level of gene expression, and that the therapeutic effect of antipsychotic medications may, in part, be mediated by the normalization of some of these alterations in gene expression. Although the cause of the altered levels of these transcripts remains to be determined, the high metabolic demands placed on neurons by the processes involved in synaptic communication suggest that these changes are related to the synaptic abnormalities present in the PFC in schizophrenia.

Interestingly, a similar microarray analysis of dorsal PFC reported significantly reduced expression of five genes expressed predominantly by glia (Hakak et al. 2001). Reports of decreased glial cell densities in schizophrenia are also beginning to emerge (Cotter et al. 2001). Recent evidence suggests that glial cells may play an active role in synaptic function, clearance of extracellular ions and transmitters, and neuronal metabolism (Bezzi and Volterra 2001). Thus, future analysis of a potential role of glial abnormalities in schizophrenia, particularly in relation to synaptic function, warrants attention.

## *GABA Neurotransmission*

These alterations in synaptic machinery and metabolism appear to be most prominent in certain neuronal populations and circuits. For example, disturbances in inhibitory (GABA) neurons appear to contribute to the dysfunction of PFC circuitry in schizophrenia. Markers of the synthesis, release, and reuptake of GABA have been reported to be decreased in schizophrenia subjects (Lewis et al. 1999). Furthermore, decreased expression of mRNAs encoding glutamate decarboxylase ($GAD_{67}$), a synthesizing enzyme for GABA, and the GABA membrane transporter (GAT-1), which is responsible for the reuptake of GABA into the nerve terminal, has been found in the PFC of subjects with schizophrenia (Volk and Lewis 2002). In addition, reduced $GAD_{67}$ mRNA expression has been found in both hemispheres of the PFC and in the temporal cortex, suggesting that dysfunction of inhibitory circuitry in cortical association regions may be a common feature of schizophrenia. In contrast, inhibitory markers appear to be undisturbed in subjects with major depressive disorder, with or without psychotic features, indicating that alterations in GABA neurotransmission may be relatively specific to the diagnosis of schizophrenia, or at least not primarily associated with either depression or psychosis. Finally, treatment with antipsychotic medication does not appear to contribute to the reduction in GABA markers in schizophrenia (Volk and Lewis 2002).

Cortical inhibitory neurons constitute distinct subpopulations of neurons that exhibit specialized anatomical, biochemical, and electrophysiological properties reflecting different roles in the regulation of neuronal activity (Lewis et al. 1999). For example, chandelier neurons provide axon terminals that exclusively target pyramidal neuron axon initial segments, the site of action potential generation. Chandelier neurons also express the calcium-binding protein parvalbumin and demonstrate a fast-spiking, nonadapting firing pattern. These characteristics suggest that chandelier neurons have the ability to powerfully regulate the output of pyramidal neurons, the principal excitatory neurons of the cortex. In contrast, double bouquet neurons, which express the calcium-binding proteins calbindin or calretinin and which exhibit a regular-spiking, adaptive firing pattern, provide axon terminals that target more distal dendritic sites on pyramidal cells or other GABA neurons. Thus, knowledge of whether disturbances in GABA neurotransmission are predominantly localized to individual subpopulations of inhibitory neurons may provide insight into the molecular basis for dysfunction of the PFC in schizophrenia.

Interestingly, alterations in inhibitory neurotransmission in schizophrenia appear to be most prominent in a subset of PFC GABA neurons that includes chandelier cells (Figure 3–1). For example, $GAD_{67}$ and GAT-1 mRNA expression levels are actually normal in the majority of inhibitory neurons but are undetectable in a subset of neurons in the same cortical layers where chandelier cells are located, and the protein level of GAT-1 appears to be selectively reduced in the axon terminals of chandelier cells (Volk and Lewis 2002). Furthermore, the protein and mRNA expression levels of parvalbumin, but not calretinin, are reduced in subjects with schizophrenia, suggesting that the affected subset of neurons probably includes chandelier cells, but not double bouquet cells. Interestingly, chandelier neurons appear to play a critical role in synchronizing the activity of pyramidal neurons and enforcing temporal precision for the integration of excitatory inputs to pyramidal neurons (Cobb et al. 1995; Pouille and Scanziani 2001). Thus, alterations in chandelier cells may be particularly relevant to the working memory dysfunction present in individuals with schizophrenia (see Goldman-Rakic 1994; Lewis and Lieberman 2000).

The consequences of schizophrenia-related disturbances in chandelier neurons can be further understood through studies of postsynaptic $GABA_A$ receptors, since changes in extracellular GABA levels appear to result in compensatory changes in $GABA_A$ receptors. Analyses of the mRNA expression of the major $GABA_A$ receptor subunits in the PFC in subjects with schizophrenia have also failed to find differences in schizophrenia, although some studies have reported increased expression of the $\alpha_1$ subunit mRNA and a reduction in the short isoform of the $\gamma_2$ subunit (see Volk and Lewis 2002 for review). However, studies of mRNA expression, which is confined to the cell soma, do not allow an analysis of differences in the distinctive subcellular distributions of different $GABA_A$ receptor subunits present in individual neurons. Interestingly, in the superficial layers of human cerebral cortex, the $\alpha_2$ subunit of the $GABA_A$ receptor is prominently localized to pyramidal neuron axon initial segment (AIS) (Loup et al. 1998). A recent study found an increase in the density of pyramidal neuron AIS immunoreactive for the $\alpha_2$ subunit in the

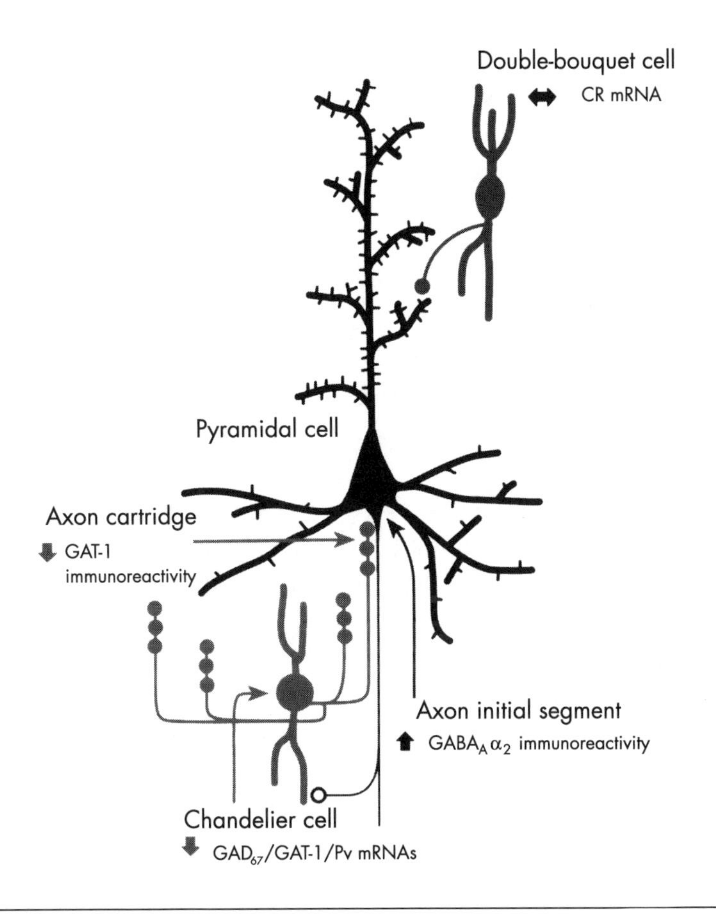


**Figure 3–1.** Schematic diagram illustrating the alterations in a subpopulation of γ-aminobutyric acid (GABA) neurons in the prefrontal cortex of subjects with schizophrenia.
CR = calretinin; $GABA_A$ = γ-aminobutyric acid receptor A; $GAD_{67}$ = 67-kDa isoform of glutamic acid decarboxylase; GAT-1 = GABA membrane transporter; PV = parvalbumin. See text for details.

PFC of subjects with schizophrenia but not in subjects with major depression, and the density of $\alpha_2$-labeled AIS was inversely correlated to the density of GAT-1–labeled cartridges in subjects with schizophrenia (Volk et al. 2002). Thus, $GABA_A$ receptors appear to be up-regulated at pyramidal neuron AIS in response to deficient GABA neurotransmission at chandelier axon terminals in schizophrenia.

The pathophysiological mechanism(s) that results in selective disturbances in chandelier neurons, but not in the majority of cortical inhibitory neurons, is currently not understood. Interestingly, although calretinin is expressed in human frontal cortex prior to birth, parvalbumin does not appear until 3–6 months of age, suggesting that the absence of this neuroprotective calcium-binding protein may create a neonatal window of vulnerability to neurotoxic events selectively in the cells that normally would eventually express parvalbumin (see Reynolds and Beasley 2001 for discussion). Alternatively, inhibitory neurotransmission may be reduced in chandelier neurons in response to a reduction in their excitatory inputs. For example, morphological changes in PFC layer 3 pyramidal neurons and reduced neuronal number in the mediodorsal nucleus of the thalamus have been reported in subjects with schizophrenia (Lewis 2000). Interestingly, the dendrites of parvalbumin-containing cells are preferentially targeted by axon terminals from neighboring pyramidal neurons and receive inputs from the mediodorsal nucleus of the thalamus (see Melchitzky and Lewis, in press). However, experimental studies of the effects of reduced excitatory input to chandelier neurons have not been reported.

Costa and colleagues (2001) have proposed a novel hypothesis that a deficiency of reelin, an extracellular matrix protein, may contribute to alterations in inhibitory neurotransmission in the PFC in schizophrenia. Their studies have revealed decreases in reelin protein and mRNA expression in multiple brain regions, including prefrontal and temporal cortex, hippocampus, caudate, and cerebellum. Interestingly, the reelin haploinsufficient mouse, which expresses approximately half of the normal levels of reelin protein and mRNA, has remarkable similarities to observations in the postmortem PFC of subjects with schizophrenia.

### *Glutamate Neurotransmission*

Several lines of evidence from clinical studies suggest that glutamatergic neurotransmission may also be deficient in schizophrenia. For example, phenylcyclidine (PCP) and ketamine, noncompetitive antagonists of the *N*-methyl-D-aspartate (NMDA) glutamate receptor, may transiently induce in control subjects positive and negative symptoms and some cognitive deficits similar to those present in schizophrenia (for review, see Goff and Coyle 2001). These drug-induced symptoms occur in adults, but rarely in children, an observation suggested to mimic the delayed age at onset of schizophrenia. Furthermore, ketamine has been reported to exacerbate positive and negative symptoms and cognitive dysfunction in subjects with schizophrenia, and this exacerbation can be reversed by treatment with clozapine (Malhotra et al. 1997). Conversely, treatment with glycine, which facilitates NMDA receptor function by binding to a modulatory site, and D-cycloserine, a selective partial agonist at the glycine modulatory site, has been reported to improve positive and negative symptoms and cognitive dysfunction in subjects with schizophrenia (Goff and Coyle 2001). Together, these clinical observations suggest that NMDA receptors may be hypofunctional in individuals with schizophrenia.

Multiple mechanisms have been proposed by which NMDA receptor hypofunction may be involved in the pathophysiology of schizophrenia. For example, chronic administration of NMDA receptor antagonists to rats and monkeys results in a reduction in cortical DA neurotransmission, particularly in the PFC, but an increase in subcortical DA activity (Jentsch and Roth 1999). These observations may be consistent with a current view of the DA hypothesis of schizophrenia (see below), which posits that deficient PFC activity of DA contributes to the cognitive features of the disorder, whereas hyperfunctional subcortical DA may be critical for psychosis (Davis et al. 1991). In addition, Olney and Farber (1995) hypothesized that hypofunction of NMDA receptors in schizophrenia primarily is manifested by reduced NMDA receptor–mediated activation of GABA neurons, and, consequently, reduced inhibition of excitatory neurons. Disinhibited excitatory neurons then release excessive glutamate, which overactivates other glutamate receptor subtypes, resulting in psychotic and cognitive dysfunction as well as neuronal death in subjects with schizophrenia. In addition, in rats treated with NMDA receptor antagonists, neurodegenerative consequences do not appear until after puberty, similar to the delayed onset of schizophrenia.

However, NMDA receptors are expressed by both pyramidal neurons and inhibitory neurons (Huntley et al. 1997), and a selective alteration in NMDA receptors in cortical inhibitory neurons in schizophrenia has not yet been reported.

In contrast to the clinical evidence and experimental animal models that support the hypothesis of NMDA receptor hypofunction in schizophrenia, genetic studies have failed to find evidence of linkage between schizophrenia and polymorphisms of the NMDAR1 subunit, an obligate subunit in NMDA receptors, or with any of the multiple metabotropic glutamate receptor subunits (see Volk and Lewis 2002 for review). Furthermore, neither radiolabeled-ligand binding studies nor mRNA expression studies in postmortem brain have provided consistent evidence of altered NMDA receptor expression in schizophrenia. However, these studies lacked the sensitivity to determine whether NMDA receptors are selectively altered in inhibitory neurons but not in pyramidal neurons, as implied by the Olney and Farber (1995) hypothesis. Studies of the other major types of ionotropic glutamate receptors, AMPA and kainate, and of metabotropic glutamate receptors in schizophrenia have also generally yielded negative findings, although several studies have reported decreased mRNA and protein expression of the AMPA receptor GluR1 and GluR2 subunits in the hippocampus and medial temporal lobe (Meador-Woodruff and Healy 2000). Interestingly, the NMDA receptor requires an initial depolarization event, such as that caused by activation of AMPA receptors located in the same synapse, to remove the voltage-dependent $Mg^{++}$ blockade of the NMDA receptor $Na^{+}$ channel. Thus, the hypothesized NMDA receptor hypofunction in schizophrenia may, at least in theory, result from reduced AMPA-mediated depolarization in hippocampus and medial temporal lobe.

Although current clinical observations and the use of animal models have built an intriguing case in support of NMDA receptor hypofunction in schizophrenia, direct evidence of disturbances in NMDA receptor expression in the brains of subjects with schizophrenia remains elusive.

## *Dopamine Neurotransmission*

Our understanding of the role of DA in the pathophysiology of schizophrenia continues to evolve. The strong correlation between the clinical efficacy of antipsychotics and their ability to block the DA $D_2$ receptor, and the tendency for DA antagonists to induce psychosis in healthy volunteers and to exacerbate psychotic symptoms in schizophrenic patients, led to the hypothesis that schizophrenia involved a functional excess of DA neurotransmission in the mesencephalic projections to the limbic striatum (Lewis and Lieberman 2000). Increased levels of $D_2$ receptors in the striatum of schizophrenic patients, in both postmortem and positron emission tomography studies, appeared to support this hypothesis. These findings, however, were not consistent, and antipsychotic drug treatment as a potential cause of the findings was difficult to exclude. Furthermore, the efficacy of atypical antipsychotics, acting on multiple neurotransmitter systems, as well as the failure of antipsychotic drugs to substantially improve the treatment of negative and cognitive symptoms, challenged the original DA hypothesis (Lewis and Lieberman 2000).

Recently, accumulating evidence of multiple DA receptor types, complex interactions of the dopaminergic system with other neurotransmitter systems, and variable DA regulation in different brain regions support a more complex DA hypothesis (Grace 1991; Olney and Farber 1995). Abnormal presynaptic storage, release, reuptake, and metabolic mechanisms in DA mesolimbic systems have been observed in schizophrenia. Furthermore, aberrant presynaptic regulation of DA may cause sensitization and/or oxidative stress and may directly contribute to the pathophysiology of schizophrenia (Lewis and Lieberman 2000). In particular, convergent lines of evidence suggest that schizophrenia may be associated with a hypodopaminergic state in the dorsal PFC. For example, deficits in working memory are a common feature of schizophrenia, and this ability depends on the optimal level of DA stimulation of $D_1$ receptors in the dorsal PFC. Interestingly, a recent postmortem study revealed a reduced density of DA axons in layer 6 of the PFC in schizophrenic subjects (Akil et al. 1999). In addition, the DA $D_1$ receptor has been reported to be upregulated in the PFC of schizophrenic patients, and the strong association with poor performance on working memory tasks suggests that deficient DA activity in the PFC may contribute to the cognitive deficits of the illness (Abi-Dargham et al. 2002). Interestingly, reduced levels of DARPP-32, a critical regulatory phosphoprotein involved in DA activity, have been found in the dorsal PFC of postmortem brains of schizophrenic patients (Albert et al. 2002). These findings, in concert with observations of an association between genetic variations in PFC DA metabolism and schizophrenia (see the section "Molecular Systems and Schizophrenia" earlier in this chapter), support the hypothesis that deficient PFC DA activity contributes to the cognitive deficits of schizophrenia.

## Implications for Understanding the Etiopathogenesis of Schizophrenia

As reviewed in this chapter, both genetic and environmental factors appear to contribute to the neurobiological abnormalities present in schizophrenia. Evidence from numerous studies suggests that interactions between genes and environment may lead to complex alterations in molecular systems of the brain, all of which are manifest as the phenotypically variable illness we call schizophrenia. Consistent with other non-Mendelian disorders, schizophrenia appears neither entirely predestined in the genome nor absolutely shaped by the environment. Our understanding as to how these factors operate independently and influence each other continues to evolve, driven by novel data, creative models, and comparisons to other illnesses.

Although rare Mendelian forms of schizophrenia may exist, most affected individuals probably carry some combination (and the total number of such combinations could be large) of allelic variations in a variety of genes, such as those for COMT, $\alpha_7$ nAChR, RGS4 and dysbindin, each of which contributes some liability for different phenotypic features of the illness. In addition, a gene associated with glutamate receptor activity, *neuregulin 1,* has recently been identified as a strong candidate gene for schizophrenia (Stefansson et al. 2002). Furthermore, such a combination of inherited factors would also likely contribute to secondary alterations in the expression of other genes whose protein products play an essential role in the types of brain functions that are disturbed in schizophrenia (Mirnics and Lewis 2001).

The altered patterns of gene expression may be influenced, or perhaps even triggered, by adverse environmental events occurring during particular sensitive periods of development, such as the prenatal, perinatal, and adolescence time frames, when rapid changes take place in different aspects of brain structure and function. Thus, individual differences in the manifestation (or lack) of disease characteristics in response to a particular environmental event would depend on the susceptibility conferred by the allelic variations associated with schizophrenia.

The combination of functional disturbances related to inherited factors and the altered patterns of gene expression due to the interaction of genetic and environmental factors may have a cumulative effect producing additional disturbances in later developmental processes (Lewis and Levitt 2002). It may be that these downstream consequences in specific brain circuits translate subtle disturbances in motor, cognitive, and social function evident early in life into the presence of the full clinical syndrome later in life.

Mirnics and colleagues (2001a) have proposed a synaptic-neurodevelopmental model of schizophrenia that encompasses genetic, developmental, and biological features of the illness. The model suggests that inheritance of some combination of allelic variants in susceptibility genes, such as the *RGS4* and *dysbindin* genes and *neuregulin 1,* produce deficits in the efficacy of synaptic transmission. These inherited liabilities in turn induce deficient expression of other synapse associated genes, as is seen in mouse gene knock-out models of specific genes regulating presynaptic machinery, which further contribute to impaired synaptic function. Although the synaptic dysfunction is present in early life, the normal exuberant synaptic connections in the cortex during early childhood compensate for the impaired synaptic function. During adolescence, the normal reduction in synaptic number reveals the functional consequences of the underlying synaptic abnormalities, giving rise to the clinical features of schizophrenia. Furthermore, the dysfunctional synapses may be excessively pruned, further compounding the functional deficit. Although incomplete, this model of the etiopathogenesis of schizophrenia is representative of current theories that attempt to explain complex disease mechanisms by integrating the potential effects of genetic, biological, and environmental risk factors.

These types of approaches to an integrated understanding of schizophrenia will likely influence future directions for treatment. The discovery of altered patterns of gene expression in schizophrenia may reveal novel molecular targets for treatment interventions. Advances in technology may lead to future pharmacogenomic treatments that target specific heritable abnormalities, prior to the development of symptoms, perhaps even in individuals determined to be at risk for illness. A great challenge facing the treatment of schizophrenia is the complexity of the disorder. The multiple etiologic factors, which appear to contribute to a constellation of clinical features, may represent both distinct and common treatment targets, prior to and throughout the course of illness.

Clearly, an integrated comprehension of the molecular neurobiology of schizophrenia rests on a number of factors: 1) the identification of the full spectrum of genetic liabilities for the illness; 2) the development of tractable model systems that permit testing of the pathophysiological processes that may link specific genetic and environmental factors with the molecular systems that are altered in schizophrenia; and 3) the establishment of the relationships between a given constellation of genetic and molecular features and phenotypes of the illness. A synthesis of molecular, genetic, and environmental factors represents the future of a neurobiological understanding of schizophrenia.

## References

Abi-Dargham A, Mawlawi O, Lombardo I, et al: Prefrontal dopamine D1 receptors and working memory in schizophrenia. J Neurosci 22: 3708–3719, 2002

Akil M, Pierri JN, Whitehead RE, et al: Lamina-specific alteration in the dopamine innervation of the prefrontal cortex in schizophrenic subjects. Am J Psychiatry 156:1580–1589, 1999

Albert KA, Hemmings HC, Adamo AIB, et al: Evidence for a decreased DARPP-32 in the prefrontal cortex of patients with schizophrenia. Arch Gen Psychiatry 59:705–712, 2002

Berman DM, Kozasa T, Gilman AG: The GTPase-activating protein RGS4 stabilizes the transition state for nucleotide hydrolysis. J Biol Chem 271:27209–27212, 1996

Bezzi P, Volterra A: A neuron-glia signalling network in the active brain. Curr Opin Neurobiol 11:387–394, 2001

Braff DL, Geyer MA: Sensorimotor gating and the neurobiology of schizophrenia: human and animal model studies, in Schizophrenia: Scientific Progress. Edited by Schulz SC, Tamminga CA. New York, Oxford University Press, 1989, pp 124–136

Breese CR, Lee MJ, Adams CE, et al: Abnormal regulation of high affinity nicotinic receptors in subjects with schizophrenia. Neuropsychopharmacology 23:351–364, 2000

Brown AS, Susser ES: In utero infection and adult schizophrenia. Ment Retard Dev Disabil Res Rev 8:51–57, 2002

Brown AS, Schaefer CA, Wyatt RJ, et al: Paternal age and risk of schizophrenia in adult offspring. Am J Psychiatry 159:1528–1533, 2002

Brzustowicz LM, Hodgkinson KA, Chow EWC, et al: Location of a major susceptibility locus for familial schizophrenia on chromosome 1q21–q22. Science 288:678–682, 2000

Cannon M, Jones PB, Murray RM: Obstetric complications and schizophrenia: historical and meta-analytic review. Am J Psychiatry 159: 1080–1092, 2002

Chen CH, Lee YR, Wei FC, et al: Association study of NlaIII and MspI genetic polymorphisms of catechol-O-methyltransferase gene and susceptibility to schizophrenia. Biol Psychiatry 41:985–987, 1997

Chowdari KV, Mirnics K, Semwal P, et al: Association and linkage analyses of RGS4 polymorphisms in schizophrenia. Hum Mol Gen 11: 1373–1380, 2002

Cobb SR, Buhl EH, Halasy K, et al: Synchronization of neuronal activity in hippocampus by individual GABAergic interneurons. Nature 378: 75–78, 1995

Costa E, Davis J, Grayson DR, et al: Dendritic spine hypoplasticity and downregulation of reelin and GABAergic tone in schizophrenia vulnerability. Neurobiol Dis 8:723–742, 2001

Cotter DR, Pariante CM, Everall IP: Glial cell abnormalities in major psychiatric disorders: the evidence and implications. Brain Res Bull 55:585–595, 2001

Daniels JK, Williams NM, Williams J, et al: No evidence for allelic association between schizophrenia and a polymorphism determining high or low catechol O-methyltransferase activity. Am J Psychiatry 153:268–270, 1996

Davis KL, Kahn RS, Ko G, et al: Dopamine in schizophrenia: a review and reconceptualization. Am J Psychiatry 148:1474–1486, 1991

De Vries L, Zheng B, Fischer T, et al: The regulator of G protein signaling family. Annu Rev Pharmacol Toxicol 40:235–271, 2000

Egan MF, Goldberg TE, Kolachana BS, et al: Effect of COMT Val 108/158 Met genotype on frontal lobe function and risk for schizophrenia. Proc Natl Acad Sci U S A 98:6917–6922, 2001

Feinberg I: Schizophrenia: caused by a fault in programmed synaptic elimination during adolescence? J Psychiatry Res 17:319–334, 1982

Freedman R, Coon H, Myles-Worsley M, et al: Linkage of a neurophysiological deficit in schizophrenia to a chromosome 15 locus. Proc Natl Acad Sci U S A 94:587–592, 1997

Freedman R, Leonard S, Gault JM, et al: Linkage disequillibrium for schizophrenia at the chromosome 15q13–14 locus of the α7-nicotinic acetylcholine receptor subunit gene (CHRNA7). Am J Med Genet 105:20–22, 2001

Goff DC, Coyle JT: The emerging role of glutamate in the pathophysiology and treatment of schizophrenia. Am J Psychiatry 158:1367–1377, 2001

Gogos JA, Morgan M, Luine V, et al: Catechol-O-methyltransferase-deficient mice exhibit sexually dimorphic changes in catecholamine levels and behavior. Proc Natl Acad Sci U S A 95:9991–9996, 1998

Goldman-Rakic PS: Working memory dysfunction in schizophrenia. J Neuropsychiatry 6:348–357, 1994

Gottesman II: Schizophrenia Genesis: The Origins of Madness. New York, WH Freeman, 1991

Grace AA: Phasic versus tonic dopamine release and the modulation of dopamine system responsivity: a hypothesis for the etiology of schizophrenia. Neuroscience 41:1–24, 1991

Green MF: What are the functional consequences of neurocognitive deficits in schizophrenia? Am J Psychiatry 153:321–330, 1996

Hakak Y, Walker JR, Li C, et al: Genomewide expression analysis reveals dysregulation of myelination-related genes in chronic schizophrenia. Proc Natl Acad Sci U S A 98:4746–4751, 2001

Harrison PJ, Lewis DA: Neuropathology in schizophrenia, in Schizophrenia. Edited by Hirsch S, Weinberger DR. Oxford, UK, Blackwell Science Ltd (in press)

Huntley G, Vickers J, Morrison J: Quantitative localization of NMDAR1 receptor subunit immunoreactivity in inferotemporal and prefrontal association cortices of monkey and human. Brain Res 749: 245–262, 1997

Jentsch JD, Roth RH: The neuropsychopharmacology of phencyclidine: from NMDA receptor hypofunction to the dopamine hypothesis of schizophrenia. Neuropsychopharmacology 20:201–225, 1999

Jones PB, Rantakallio P, Hartikainen A-L, et al: Schizophrenia as a long-term outcome of pregnancy, delivery, and perinatal complications: a 28-year follow-up of the 1966 North Finland general population birth cohort. Am J Psychiatry 155:355–364, 1998

Karayiorgou M, Morris MA, Morrow B, et al: Schizophrenia susceptibility associated with interstitial deletions of chromosome 22q11. Proc Natl Acad Sci U S A 92:7612–7616, 1995

Karayiorgou M, Gogos JA, Galke BL, et al: Identification of sequence variants and analysis of the role of the catechol-O-methyltransferase gene in schizophrenia susceptibility. Biol Psychiatry 43:425–431, 1998

Keshavan MS, Hogarty GE: Brain maturational processes and delayed onset in schizophrenia. Dev Psychopathol 11:525–543, 1999

Leonard S, Adams C, Breese CR, et al: Nicotinic receptor function in schizophrenia. Schizophr Bull 22:431–445, 1996

Lewis DA: Is there a neuropathology of schizophrenia? The Neuroscientist 6:208–218, 2000

Lewis DA: Neural circuitry approaches to understanding the pathophysiology of schizophrenia, in Neuropsychopharmacology: The Fifth Generation of Progress. Edited by Davis KL, Charney DS, Coyle JT, et al. Philadelphia, PA, Lippincott Williams & Wilkins, 2002, pp 729–743

Lewis DA, Levitt P: Schizophrenia as a disorder of neurodevelopment. Annu Rev Neurosci 25:409–432, 2002

Lewis DA, Lieberman JA: Catching up on schizophrenia: natural history and neurobiology. Neuron 28:325–334, 2000

Lewis DA, Pierri JN, Volk DW, et al: Altered GABA neurotransmission and prefrontal cortical dysfunction in schizophrenia. Biol Psychiatry 46:616–626, 1999

Liou YJ, Tsai SJ, Hong CJ, et al: Association analysis of a functional catechol-O-methyltransferase gene polymorphism in schizophrenic patients in Taiwan. Neuropsychobiology 43:11–14, 2001

Loup F, Weinmann O, Yonekawa Y, et al: A highly sensitive immunoflourescence procedure for analyzing the subcellular distribution of $GABA_A$ receptor subunits in the human brain. J Histochem Cytochem 46:1129–1139, 1998

Malaspina D, Harlap S, Fennig S, et al: Advancing paternal age and the risk of schizophrenia. Arch Gen Psychiatry 58:361–367, 2001

Malhotra AK, Adler CM, Kennison SD, et al: Clozapine blunts N-methyl-D-aspartate antagonist–induced psychosis: a study with ketamine. Biol Psychiatry 42:664–668, 1997

Marenco S, Weinberger DR: The neurodevelopmental hypothesis of schizophrenia: following a trail of evidence from cradle to grave. Dev Psychopathol 12:501–527, 2000

McCarley RW, Wible CG, Frumin M, et al: MRI anatomy of schizophrenia. Biol Psychiatry 45:1099–1119, 1999

Meador-Woodruff JH, Healy DJ: Glutamate receptor expression in schizophrenic brain. Brain Res Rev 31:288–294, 2000

Melchitzky DS, Lewis DA: Preferential targeting of parvalbumin interneurons by local axon terminals of supragranular pyramidal neurons in monkey prefrontal cortex. Cereb Cortex (in press)

Middleton FA, Mirnics K, Pierri JN, et al: Gene expression profiling reveals alterations of specific metabolic pathways in schizophrenia. J Neurosci 22:2718–2729, 2002

Mirnics K, Lewis DA: Genes and subtypes of schizophrenia. Trends Mol Med 7:281–283, 2001

Mirnics K, Middleton FA, Marquez A, et al: Molecular characterization of schizophrenia viewed by microarray analysis of gene expression in prefrontal cortex. Neuron 28:53–67, 2000

Mirnics K, Middleton FA, Lewis DA, et al: Analysis of complex brain disorders with gene expression microarrays: schizophrenia as a disease of the synapse. Trends Neurosci 24:479–486, 2001a

Mirnics K, Middleton FA, Stanwood GD, et al: Disease-specific changes in regulator of G-protein signaling 4 (RGS4) expression in schizophrenia. Mol Psychiatry 6:293–301, 2001b

Olincy A, Young DA, Freedman R: Increased levels of the nicotine metabolite cotinine in schizophrenic smokers compared to other smokers. Biol Psychiatry 42:1–5, 1997

Olney JW, Farber NB: Glutamate receptor dysfunction and schizophrenia. Arch Gen Psychiatry 52:998–1007, 1995

Pedersen CB, Mortensen PB: Evidence of a dose-response relationship between urbanicity during upbringing and schizophrenia risk. Arch Gen Psychiatry 58:1039–1046, 2001

Pierri JN, Volk CLE, Auh S, et al: Decreased somal size of deep layer 3 pyramidal neurons in the prefrontal cortex in subjects with schizophrenia. Arch Gen Psychiatry 58:466–473, 2001

Pouille F, Scanziani M: Enforcement of temporal fidelity in pyramidal cells by somatic feed-forward inhibition. Science 293:1159–1163, 2001

Rajkowska G, Selemon LD, Goldman-Rakic PS: Neuronal and glial somal size in the prefrontal cortex: a postmortem morphometric study of schizophrenia and Huntington disease. Arch Gen Psychiatry 55: 215–224, 1998

Reynolds GP, Beasley CL: GABAergic neuronal subtypes in the human frontal cortex—development and deficits in schizophrenia. J Chem Neuroanat 22:95–100, 2001

Riley BP, McGuffin P: Linkage and associated studies of schizophrenia. Am J Med Genet 97:23–44, 2000

Selemon LD, Goldman-Rakic PS: The reduced neuropil hypothesis: a circuit-based model of schizophrenia. Biol Psychiatry 45:17–25, 1999

Shenton ME, Dickey CC, Frumin M, et al: A review of MRI findings in schizophrenia. Schizophr Res 49:1–52, 2001

Stefansson H, Sigurdsson E, Steinthorsdottir V, et al: Neuregulin 1 and susceptibility to schizophrenia. Am J Hum Genet 71:877–892, 2002

Straub RE, Jiang Y, MacLean CJ, et al: Genetic variation in the 6p22.3 gene DTNBP1, the human ortholog of the mouse dysbindin gene, is associated with schizophrenia. Am J Hum Genet 71:337–348, 2002

van Erp TGM, Saleh PA, Rosso IM, et al: Contributions of genetic risk and fetal hypoxia to hippocampal volume in patients with schizophrenia of schizoaffective disorder, their unaffected siblings, and healthy unrelated volunteers. Am J Psychiatry 159:1514–1520, 2002

Volk DW, Lewis DA: Impaired prefrontal inhibition in schizophrenia: relevance for cognitive dysfunction. Physiol Behav 77:537–543, 2002

Volk DW, Lewis DA: Schizophrenia, in The Molecular and Genetic Basis of Neurologic and Psychiatric Disease. Edited by Rosenberg R, Prusiner S, DiMauro S, et al. Woburn, MA, Butterworth Heineman (in press)

Volk DW, Pierri JN, Fritschy J-M, et al: Reciprocal alterations in pre- and postsynaptic inhibitory markers at chandelier cell inputs to pyramidal neurons in schizophrenia. Cereb Cortex 12:1063–1070, 2002

Weiland S, Bertrand D, Leonard S: Neuronal nicotinic acetylcholine receptors: from the gene to the disease. Behav Brain Res 113:43–56, 2000

Weinberger DR: Implications of normal brain development for the pathogenesis of schizophrenia. Arch Gen Psychiatry 44:660–669, 1987

Weinberger DR, Egan MF, Bertolino A, et al: Prefrontal neurons and the genetics of schizophrenia. Biol Psychiatry 50:825–844, 2001
Woods BT: Is schizophrenia a progressive neurodevelopmental disorder? Toward a unitary pathogenetic mechanism. Am J Psychiatry 155:1661–1670, 1999

Chapter 4

# Molecular Mechanisms of Drug Addiction

*Eric J. Nestler, M.D., Ph.D.*

Drug addiction is a behaviorally defined syndrome, characterized by loss of control over drug intake and compulsive drug taking despite horrendous adverse consequences. These behavioral abnormalities generally develop gradually and progressively during a course of repeated exposure to a drug of abuse and can persist for months or years after discontinuation of drug use. As a result, drug addiction can be considered a form of drug-induced neural plasticity (Nestler et al. 1993). The stability of the behavioral abnormalities that define addiction has led several groups to consider a role for gene expression in this process. According to this view, repeated exposure to a drug of abuse alters the amounts, and even the types, of genes expressed in specific brain regions. Such altered expression of genes then mediates altered function of individual neurons and the larger neural circuits within which the neurons operate. Ultimately, such neural circuit changes underlie the behavioral abnormalities seen in individuals with drug addictions (Berke and Hyman 2000; Nestler 2001; Nestler et al. 1993).

There are many mechanisms by which repeated exposure to a drug of abuse could alter gene expression in the brain. These include altered rates of transcription of genes, altered processing of primary RNA transcripts into mature mRNAs, altered translation of these mRNAs into proteins, altered processing of proteins,

Preparation of this review was supported by grants from the National Institute on Drug Abuse.

and altered trafficking of mature proteins to their intracellular sites of action (see Nestler et al. 2001b). Of all these mechanisms, the best understood, and the one that has received most study to date, is the regulation of gene transcription. According to this scheme, illustrated in Figure 4–1, drug perturbation of synaptic transmission causes changes in numerous intracellular signaling pathways, which eventually signal to the cell nucleus, where specific proteins, called *transcription factors,* are altered. Transcription factors bind to short sequences of DNA located in the regulatory regions of genes and thereby control the rate of gene transcription. Over the past decade, drugs of abuse have been shown to alter many types of transcription factors in a variety of brain regions (see Berke and Hyman 2000; Nestler 2001; O'Donovan et al. 1999 for discussion).

This review focuses on two transcription factors, CREB (cAMP response element binding protein) and ΔFosB, which our laboratory has investigated. It is important to state at the outset that these particular proteins represent just a small part of the overall plasticity that drugs of abuse induce in the brain. Nevertheless, consideration of these proteins serves as an illustration of how reductionistic changes (alterations in a single transcription factor in a particular neuronal cell type in the brain) can be related to something as complex as the behavioral phenotype of addiction.

## Role of CREB in Drug Addiction

CREB and related proteins were first discovered as transcription factors that mediate effects of the cAMP second messenger pathway on gene expression (Mayr and Montminy 2001). This occurs via the phosphorylation of CREB on a single serine residue, Ser133, by protein kinase A (a protein kinase activated by cAMP). Once phosphorylated, CREB dimers, bound to specific CRE (cAMP response element) sites on target genes, can interact with the basal transcriptional complex to regulate gene transcription. More recently, several other protein kinases, including $Ca^{2+}$ activated and growth factor activated kinases, have been shown to phosphorylate Ser133 of CREB and to thereby regulate the transcription of the same set of CREB-sensitive genes.

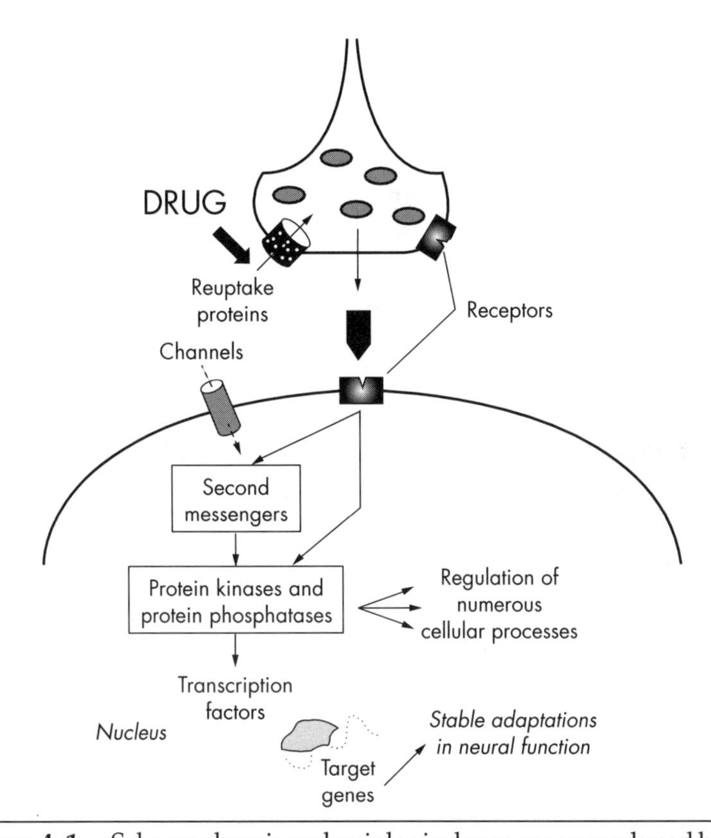


**Figure 4–1.** Scheme showing physiological responses produced by the initial effects of drugs of abuse on their acute synaptic targets.
The initial effects of drugs of abuse on their acute synaptic targets, which are located extracellularly, produce physiological responses via perturbation of postreceptor intracellular messenger pathways that mediate extracellular mechanisms. Among the physiological responses produced are alterations in gene expression mediated via drug-induced changes in transcription factors.
*Source.* Reprinted from Nestler EJ: "Molecular Neurobiology of Addiction." *American Journal on Addictions* 10:201–217, 2001. Used with permission.

The first indication that CREB mediates aspects of addiction came from studies of the locus coeruleus, the major noradrenergic nucleus in brain. This brain region normally regulates attentional states and is thought to play an important role in mediating physical opiate dependence and withdrawal (Koob et al. 1992). One mechanism responsible for opiate physical dependence is an upregulation of the cAMP pathway, which occurs in these neurons as a consequence of repeated opiate exposure (Nestler and Aghajanian 1998). Several lines of evidence directly implicate opiate-induced activation of CREB in mediating this upregulation of the cAMP pathway in the locus coeruleus (Lane-Ladd et al. 1997). This is generally consistent with the observation that mice deficient in CREB in all tissues show attentuated opiate physical dependence and withdrawal (Maldonado et al. 1996).

However, although the locus coeruleus is important for physical opiate dependence and withdrawal, it is not crucial for the effects of drugs on the brain's reward and motivational systems, which appear to be more critical for drug addiction. These latter regions include several overlapping circuits, comprising the nucleus accumbens (ventral striatum), ventral tegmental area, amygdala, hippocampus, lateral hypothalamus, and frontal cortex, to name a few. Interestingly, upregulation of the cAMP pathway and activation of CREB occur in many of these other regions as originally established for the locus coeruleus (see Berke and Hyman 2000; Carlezon et al. 1998; Cole et al. 1995; Nestler 2001; Shaw-Lutchman et al. 2002, 2003). In several of these regions, activation of CREB occurs in response not only to chronic opiate exposure but also to chronic administration of other drugs of abuse, in particular stimulants and alcohol. A leading hypothesis is that drug-induced activation of CREB in these motivation centers of the brain underlies some of the common core features of drug addiction seen clinically.

Our best understanding of the precise role played by CREB in motivational aspects of addiction concerns the nucleus accumbens. We have used herpes simplex viral (HSV) vectors to induce local changes in CREB activity within the nucleus accumbens of adult animals. One vector (HSV-CREB) encodes wild-type or normal CREB; this vector increases CRE-mediated transcription on injection into the brain. A second vector (HSV-mCREB) encodes a mutant form of CREB, called mCREB, which is identical to wild-type CREB except that it has an alanine residue substituted for Ser133. Consequently, mCREB cannot mediate transcription and, when expressed at high enough levels, can block the action of endogenous CREB and CREB-like proteins; HSV-mCREB decreases CRE-mediated transcription on injection into the brain. Using these viral vectors, we have found that increased CREB function (via HSV-CREB injection) in the nucleus accumbens decreases an animal's sensitivity to the rewarding effects of morphine and cocaine, whereas decreased CREB function (via HSV-mCREB injection) has the opposite effect (Barrot et al. 2002; Carlezon et al. 1998). These results suggest that drug-induced activation of CREB in the nucleus accumbens represents a homeostatic, negative feedback adaptation that serves to limit an individual's sensitivity to subsequent drug exposure. In this way, CREB activation could mediate a form of tolerance to a drug's rewarding effects.

However, the consequences of CREB activation in the nucleus accumbens go far beyond modulation of drug reward. Increased CREB function in this region also dampens an animal's interest for natural rewards, such as sucrose drinking, while decreased CREB function enhances sucrose drinking (Barrot et al. 2002). These results suggest that drug-induced activation of CREB in the nucleus accumbens could decrease an individual's interest in natural rewards, as is seen in many human addicts.

Moreover, exposure of an animal to stress causes a similar activation of CREB in the nucleus accumbens as seen with drugs of abuse. Accordingly, we found that increased CREB function in the nucleus accumbens decreases an animal's responsiveness to a variety of aversive or negative emotional stimuli, including anxiogenic, stressful, and nociceptive stimuli (Barrot et al. 2002; Pliakas et al. 2001). Conversely, decreased CREB function in this region increases the animal's sensitivity to these conditions. Hence, CREB activation in the nucleus accumbens would appear to mediate a more general behavioral syndrome, characterized by reduced sensitivity to any type of emotional stimulus, regardless of whether the stimulus is rewarding or aversive. In a physiological context, CREB activation in response to mild stress could be seen as part of a coping mechanism to diminish sensitivity to further stress. However, under more extreme (pathological) circumstances, CREB activation could contribute to a behavioral syndrome of emotional numbing and anhedonia, as seen both during drug withdrawal syndromes and in certain cases of depression and posttraumatic stress disorder (Nestler et al. 2002).

All of the above conclusions concerning CREB's functioning at the level of the nucleus accumbens are based on the use of viral vectors, which could produce confounding results due to the fact that all the animals receive intracranial surgery and viral infections. However, early work with inducible transgenic mice, in which CREB or mCREB is expressed with some selectivity in the nucleus accumbens, supports these conclusions. Thus, mice that overexpress CREB show reduced sensitivity to drugs of abuse and stress, whereas mice that express mCREB show increased sensitivity (Newton et al. 2002; Sakai et al. 2002). The fact that the behavioral phenotype of CREB is reproduced with a completely independent method of locally controlling CREB function in adult animals gives us greater confidence in our current working hypothesis of CREB action.

These same types of approaches are now needed to determine the role played by drug-induced activation of CREB in the other reward-related brain regions mentioned above. Early work suggests that CREB in certain of these regions—amygdala and lateral hypothalamus, for example—may subserve very different aspects of the drug addiction phenotype compared with the nucleus accumbens (Georgescu et al., in press; Jentsch et al. 2002; Josselyn et al. 2001).

As a transcription factor, CREB produces its behavioral effects via the regulation of other genes. Numerous genes are known to contain CRE sites within their promoter regions (see Mayr and Montminy 2001). However, genes that are regulated by CREB in the nucleus accumbens or elsewhere remain largely obscure. Work to date has implicated the gene for the opioid peptide dynorphin as one relevant target for CREB in nucleus accumbens. Chronic administration of cocaine or other stimulants induces dynorphin expression in this region, and this induction is dependent on CREB (Carlezon et al. 1998; Cole et al. 1995). Dynorphin is known to dampen reward mechanisms in the nucleus accumbens via activation of κ opioid receptors (Shippenberg and Rea 1997), suggesting that CREB induction of dynorphin could mediate at least some of CREB's behavioral effects in this region. Indeed, the ability of CREB activation to dampen cocaine reward, as well as responses to stress, can be blocked by κ opioid receptor antagonists (Carlezon et al. 1998; Pliakas et al. 2001), which is consistent with this model (Figure 4–2). A major focus of current research is to identify additional target genes through which CREB also acts in this and other brain regions.

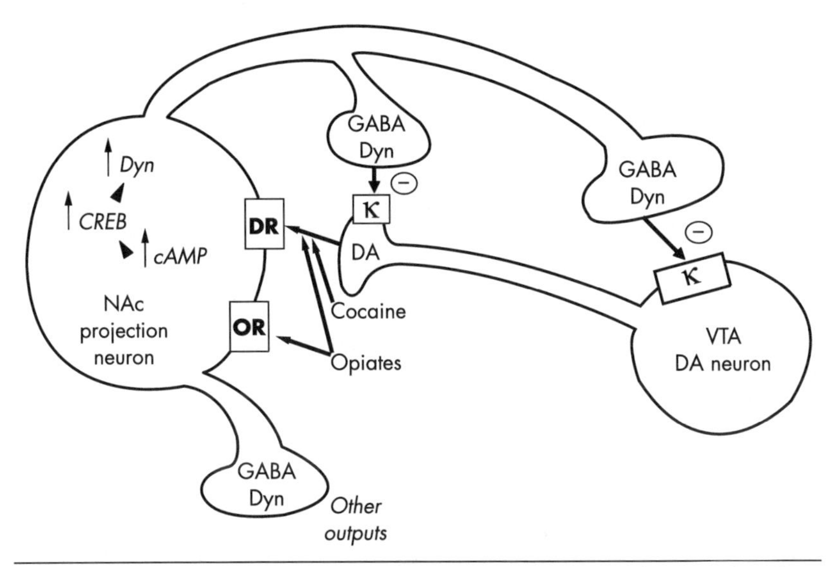


**Figure 4–2.** Regulation of CREB (cAMP response element binding protein) by drugs of abuse.

The figure shows a ventral tegmental area (VTA) dopamine (DA) neuron innervating a class of NAc GABAergic projection neuron that expressed dynorphin (dyn). Dynorphin serves a negative feedback mechanism in this circuit: dynorphin, released from terminals of the NAc neurons, acts on κ opioid receptors located on nerve terminals and cell bodies of the DA neurons to inhibit their functioning. Chronic exposure to cocaine or opiates upregulates the activity of this negative feedback loop via upregulation of the cAMP pathway, activation of CREB, and induction of dynorphin. DR = dopamine receptor; OR = opioid receptor.

*Source.* Reprinted from Nestler EJ: "Molecular Neurobiology of Addiction." *American Journal on Addictions* 10:201–217, 2001. Used with permission.

## Role of ΔFosB in Drug Addiction

ΔFosB is a member of the Fos family of transcription factors. These proteins dimerize with a Jun family member to form AP-1 (activator protein–1) transcription factor complexes, which bind to AP-1 sites present within the regulatory regions of certain genes

(Morgan and Curran 1995). Acute administration of most types of drugs of abuse causes the rapid and transient induction of several Fos and Jun proteins in the nucleus accumbens and dorsal striatum (see Nestler et al. 2001a). The transient nature of this induction is due to the fact that the mRNAs for these proteins and the proteins themselves are highly unstable. In contrast, ΔFosB is unique among these proteins in that it is induced to only a small degree in response to initial drug exposures. In addition, it is a highly stable protein. As a result, during a course of repeated drug administration, ΔFosB gradually accumulates and after a time becomes the predominant Fos protein present in the nucleus accumbens and dorsal striatum (Hope et al. 1994; Moratalla et al. 1996) (Figure 4–3). Also, because of its stability, ΔFosB, once induced, persists in these regions for weeks or months after discontinuation of drug exposure. The accumulation of ΔFosB therefore represents a novel mechanism by which chronic drug exposure can lead to changes in gene expression that persist long after drug taking ceases (Nestler et al. 2001a).

Insight into the function of ΔFosB in these regions comes from studies of inducible transgenic mice, where ΔFosB can be induced selectively within the nucleus accumbens and dorsal striatum of adult mice (Kelz et al. 1999). Studies to date indicate three major phenotypes in mice that overexpress ΔFosB. First, the animals show greater sensitivity to the behavioral effects of cocaine in locomotor activity, conditioned place preference, and self-administration assays (Kelz et al. 1999; Nestler et al. 2001a). Second, the mice seem to show enhanced incentive motivation for cocaine. This conclusion is based on studies in progressive ratio tests and in a model of relapse that involves nonreinforced responding (Nestler et al. 2001a). Third, the mice show greater sensitivity to the behavioral effects of morphine and greater responding to naturally reinforcing behaviors, including running and eating (Nestler et al. 2001a; Werme et al. 2002). Together, these findings suggest that drug-induced accumulation of ΔFosB in nucleus accumbens and dorsal striatum could be a mechanism of relatively prolonged sensitization to drug exposure. Moreover, the findings suggest that ΔFosB could be part of a sustained molecular switch, which functions to first induce and later maintain

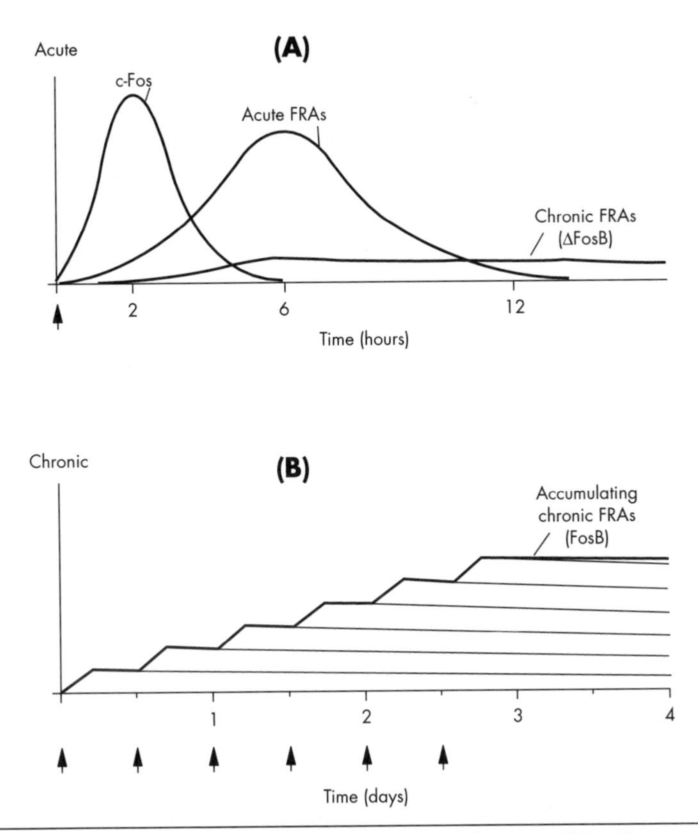


**Figure 4–3.** Scheme for the gradual accumulation of FosB versus the rapid and transient induction of acute Fos family proteins in brain.
**(A)** Several waves of Fos-like proteins are induced in neurons by acute stimuli. c-Fos is induced rapidly and degraded within several hours of the acute stimulus, whereas other "acute FRAs" (Fos-related antigens; e.g., FosB, ΔFosB, FRA-1, FRA-2) are induced somewhat later and persist somewhat longer than c-Fos. The "chronic FRAs" are biochemically modified isoforms of ΔFosB; they, too, are induced (although at low levels) following a single acute simulus but persist in brain for long periods. In a complex with Jun-like proteins, these waves of FRAs form AP-1 binding complexes with shifting composition over time. **(B)** With repeated (e.g., twice daily) stimulation, each acute stimulus induces a low level of ΔFosB. This is indicated by the lower set of overlapping lines, which indicate ΔFosB induced by each acute stimulus. The result is a gradual increase in the total levels of ΔFosB with repeated stimuli during a course of chronic treatment, as indicated by the increasing stepped line in the graph. The increasing levels of ΔFosB with repeated stimulation would result in the gradual induction of significant levels of a long-lasting AP-1 complex, which is hypothesized to underlie persisting forms of neural plasticity in the brain.
*Source.* Reprinted from Nestler EJ: "Molecular Neurobiology of Addiction." *American Journal on Addictions* 10:201–217, 2001. Used with permission.

a state of heightened incentive motivation, perhaps even compulsiveness, toward reinforced behaviors (Nestler et al. 2001a).

Recent work, in a different mouse model, has provided further support for this hypothesis. The mice in this model enable the expression of a dominant negative mutant of c-Jun, which functions as an antagonist of ΔFosB and other Fos family proteins, in nucleus accumbens and dorsal striatum. These mice show reduced sensitivity to the behavioral effects of cocaine in conditioned place preference assays (see Peakman et al., in press). Along with the findings from ΔFosB-overexpressing mice, these data suggest that FosB is both necessary and sufficient for induction of a sensitized behavioral state after drug exposure. Analysis of this mouse in other behavioral assays is under way.

As with CREB, a major interest is to identify target genes through which ΔFosB produces these behavioral effects in nucleus accumbens and dorsal striatum. Several putative target genes have been found. GluR2 is an AMPA glutamate receptor subunit. Overexpression of ΔFosB induces GluR2 in nucleus accumbens, and cocaine induction of GluR2 in this region is blocked on expression of dominant negative c-Jun (Kelz et al. 1999; Peakman et al., in press). Moreover, overexpression of GluR2 in nucleus accumbens, by use of HSV vectors, causes increased sensitivity to the behavioral effects of cocaine, indicating that induction of GluR2 could account for at least some of the ΔFosB behavioral phenotype. Since AMPA receptors containing a GluR2 subunit show reduced conductance compared with those not containing GluR2 subunits, it is possible that cocaine- and ΔFosB-mediated induction of GluR2 could account for the reduced sensitivity of nucleus accumbens neurons to glutamate that has been observed in electrophysiological studies (Thomas et al. 2001; White et al. 1995).

Another putative target for ΔFosB is the cyclin-dependent kinase Cdk5. Cdk5 is enriched in neural tissues, where it is implicated in the regulation of neuronal growth and survival. The identification of Cdk5 as a target for ΔFosB came from DNA microarray studies (Chen et al. 2000). Subsequently, chronic cocaine administration was shown to induce Cdk5 mRNA and protein expression, as well as Cdk5 catalytic activity, in nucleus accumbens and dorsal striatum (Bibb et al. 2001). Based on Cdk5's hypothesized role in neural growth, we considered the possibility that cocaine- and ΔFosB-induction of Cdk5 could be involved in cocaine's demonstrated ability to induce dendritic spines in nucleus accumbens neurons (Robinson and Kolb 1997). Indeed, infusion of a Cdk5 inhibitor directly into the nucleus accumbens was found to completely prevent the ability of cocaine to induce dendritic spines in this region (see Norrholm et al., in press). This finding raises the notion that ΔFosB may be responsible for this morphological plasticity and, moreover, that some of ΔFosB's impact on the nucleus accumbens may be more long-lasting than the protein itself.

Still other putative targets for ΔFosB have been identified. ΔFosB induces NF-κB (nuclear factor–κB), another transcription factor, in nucleus accumbens (Ang et al. 2001). Many additional targets have been identified preliminarily on DNA microarrays (Chen et al. 2000). Together, these results suggest considerable complexity to the cascades of molecular changes that ΔFosB (and drugs of abuse) induces in this region.

## Conclusion

This review focused on two transcription factors, CREB and ΔFosB, as targets for drugs of abuse. Both are induced in nucleus accumbens, among other brain regions, but seem to function very differently even in that one region (Figure 4–4). Chronic exposure to drugs of abuse causes the activation of CREB in this region, but this activation dissipates within a few days of coming off of the drug. In contrast, ΔFosB persists in brain for up to 2 months, meaning that it mediates a much more long-lived signal compared with CREB. In addition, while CREB activation seems to mediate a state of reduced reward and reduced emotional reactivity, ΔFosB accumulation mediates a state of heightened drug sensitivity and increased compulsion for rewarding behavior.

We are still in relatively early stages of identifying the many target genes in nucleus accumbens through which CREB and ΔFosB produce these behavioral effects. Moreover, we are just beginning to analyze the behavioral consequences of drug-induced

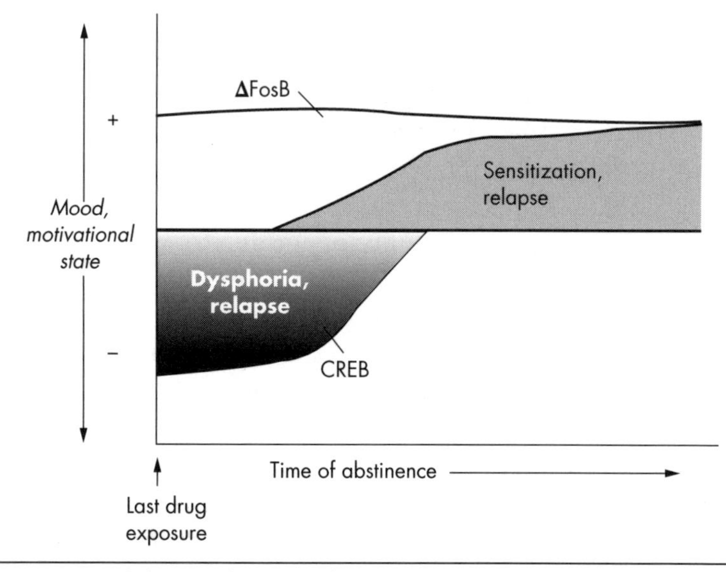


**Figure 4–4.** Scheme showing the opposite functional effects, and distinct temporal properties, of drug-induced adaptations mediated via CREB (cAMP response element binding protein) versus ΔFosB.
CREB appears to mediate an aversive or dysphoric state during early phases of withdrawal, whereas ΔFosB appears to mediate sensitization to subsequent drug exposures at more distant withdrawal times.
*Source.* Reprinted from Nestler EJ: "Molecular Neurobiology of Addiction." *American Journal on Addictions* 10:201–217, 2001. Used with permission.

activation of CREB and ΔFosB in other brain regions where these proteins are induced by drug exposure. Finally, these two proteins are among a very large number of molecular changes that drugs of abuse are known to cause in reward-related brain regions after chronic administration. Ultimately, it will be critical to integrate this growing body of information to understand how drugs, through plastic changes involving numerous proteins, induce a complex behavioral phenotype that defines the addicted state.

## References

Ang E, Chen JS, Zagouras P, et al: Induction of NFκB in nucleus accumbens by chronic cocaine administration. J Neurochem 79:221–224, 2001

Barrot M, Olivier JDA, Perrotti LI, et al: CREB activity in the nucleus accumbens shell controls gating of behavioral responses to emotional stimuli. Proc Natl Acad Sci U S A 99:11435–11440, 2002

Berke JD, Hyman SE: Addiction, dopamine, and the molecular mechanisms of memory. Neuron 25:515–532, 2000

Bibb JA, Chen JS, Taylor JR, et al: Cdk5 regulates action of chronic cocaine. Nature 410:376–380, 2001

Carlezon WA Jr, Thome J, Olson VG, et al: Regulation of cocaine reward by CREB. Science 282:2272–2275, 1998

Chen JS, Zhang YJ, Kelz MB, et al: Induction of cyclin-dependent kinase 5 in hippocampus by chronic electroconvulsive seizures: role of ΔFosB. J Neurosci 20:8965–8971, 2000

Cole RL, Konradi C, Douglass J, et al: Neuronal adaptation to amphetamine and dopamine: molecular mechanisms of prodynorphin gene regulation in rat striatum. Neuron 14:813–823, 1995

Georgescu D, Zachariou V, Barrot M, et al: Involvement of the lateral hypothalamic peptide orexin in morphine dependence and withdrawal. J Neurosci (in press)

Hope BT, Nye HE, Kelz MB, et al: Induction of a long-lasting AP-1 complex composed of altered Fos-like proteins in brain by chronic cocaine and other chronic treatments. Neuron 13:1235–1244, 1994

Jentsch JD, Olausson P, Nestler EJ, et al: Stimulation of protein kinase A activity in the rat amygdala enhances reward-related learning. Biol Psychiatry 52:111–118, 2002

Josselyn SA, Shi CJ, Carlezon WA Jr, et al: Long-term memory is facilitated by CREB overexpression in the amygdala. J Neurosci 21:2404–2412, 2001

Kelz MB, Chen JS, Carlezon WA, et al: Expression of the transcription factor ΔFosB in the brain controls sensitivity to cocaine. Nature 401: 272–276, 1999

Koob GF, Maldonado R, Stimus L: Neural substrates of opiate withdrawal. Trends Neurosci 15:186–191, 1992

Lane-Ladd SB, Pineda J, Boundy V, et al: CREB (cAMP response element binding protein) in the locus coeruleus: biochemical, physiological, and behavioral evidence for a role in opiate dependence. J Neurosci 17:7890–7901, 1997

Maldonado R, Blendy JA, Tzavara E, et al: Reduction of morphine abstinence in mice with a mutation in the gene encoding CREB. Science 273:657–659, 1996

Mayr B, Montminy M: Transcriptional regulation by the phosphorylation-dependent factor CREB. Nature Rev Mol Cell Biol 2:599–609, 2001

Moratalla R, Elibol B, Vallejo M, et al: Network-level changes in expression of inducible Fos-Jun proteins in the striatum during chronic cocaine treatment and withdrawal. Neuron 17:147–156, 1996

Morgan JI, Curran T: Immediate-early genes: ten years on. Trends Neurosci 18:66–67, 1995

Nestler EJ: Molecular basis of long-lived neural plasticity to drugs of abuse. Nature Rev Neurosci 2:119–128, 2001

Nestler EJ, Aghajanian GK: Molecular and cellular basis of addiction. Science 278:58–63, 1998

Nestler EJ, Hope BT, Widnell KL: Drug addiction: a model for the molecular basis of neural plasticity. Neuron 11:995–1006, 1993

Nestler EJ, Barrot M, Self DW: ΔFosB: a molecular switch for addiction. Proc Natl Acad Sci U S A 98:11042–11046, 2001a

Nestler EJ, Hyman SE, Malenka RC: Molecular Basis of Neuropharmacology. New York, McGraw-Hill, 2001b

Nestler EJ, Barrot M, DiLeone RJ, et al: Neurobiology of depression. Neuron 34:13–25, 2002

Newton SS, Thome J, Wallace T, et al: Inhibition of cAMP response element–binding protein or dynorphin in the nucleus accumbens produces an antidepressant-like effect. J Neurosci 22:10883–10890, 2002

Norrholm SD, Bibb JA, Nestler EJ, et al: Cocaine-induced proliferation of dendritic spines in nucleus accumbens is dependent on the activity of the neuronal kinase Cdk5. Neuroscience (in press)

O'Donovan KJ, Tourtellotte WG, Millbrandt J, et al: The EGR family of transcription-regulatory factors: progress at the interface of molecular and systems neuroscience. Trends Neurosci 22:167–173, 1999

Peakman MC, Colby C, Duman C, et al : Inducible, brain region specific expression of a dominant negative mutant of c-Jun in transgenic mice decreases sensitivity to cocaine. Brain Res (in press)

Pliakas AM, Carlson RR, Neve RL, et al: Altered responsiveness to cocaine and increased immobility in the forced swim test associated with elevated CREB expression in the nucleus accumbens. J Neurosci 21:7397–7403, 2001

Robinson TE, Kolb B: Persistent structural modifications in nucleus accumbens and prefrontal cortex neurons produced by previous experience with amphetamine. J Neurosci 17:8491–8497, 1997

Sakai N, Thome J, Chen JS, et al: Inducible and brain region specific CREB transgenic mice. Mol Pharmacol 61:1453–1464, 2002

Shaw-Lutchman TZ, Barrot M, Wallace T, et al: Regional and cellular mapping of CRE-mediated transcription during naltrexone-precipitated morphine withdrawal. J Neurosci 22:3663–3672, 2002

Shaw-Lutchman SZ, Impey S, Storm D, et al: Regulation of CRE-mediated transcription in mouse brain by amphetamine. Synapse 48:10–17, 2003

Shippenberg TS, Rea W: Sensitization to the behavioral effects of cocaine: modulation by dynorphin and kappa-opioid receptor agonists. Pharmacol Biochem Behav 57:449–455, 1997

Thomas MJ, Beurrier C, Bonci A, et al: Long-term depression in the nucleus accumbens: a neural correlate of behavioral sensitization to cocaine. Nature Neurosci 4:1217–1223, 2001

Werme M, Messer C, Olson L, et al: ΔFosB regulates wheel running. J Neurosci 22:8133–8138, 2002

White FJ, Hu XT, Zhang X-F, et al: Repeated administration of cocaine or amphetamine alters neuronal responses to glutamate in the mesoaccumbens dopamine system. J Pharmacol Exp Ther 273:445–454, 1995

Chapter 5

# Cellular Neurobiology of Severe Mood and Anxiety Disorders

## Implications for Development of Novel Therapeutics

*Todd D. Gould, M.D.*
*Neil A. Gray, B.A.*
*Husseini K. Manji, M.D., F.R.C.P.C.*

Despite the devastating impact that mood and anxiety disorders have on the lives of millions worldwide, there is still a dearth of knowledge concerning their underlying etiology and pathophysiology. The brain systems that have heretofore received the greatest attention in neurobiological studies of mood disorders have been the monoaminergic neurotransmitter systems, which are extensively distributed throughout the network of limbic, striatal, and prefrontal cortical neuronal circuits that are thought to support the behavioral and visceral manifestations of mood disorders (Drevets 2001; Manji et al. 2001a; Nestler et al. 2002). There is a growing appreciation that these disorders likely arise from the complex interaction of multiple susceptibility (and protective) genes and environmental factors, and their phenotypic expression includes not only profound mood and anxiety disturbances but also a constellation of cognitive, motoric, autonomic, endocrine, and sleep/wake abnormalities. Furthermore, with the exception of benzodiazepines, all the existing effective treatments for these disorders require chronic administration for therapeutic benefits, indicating that delayed, adaptive changes in critical neuronal circuits are much more important than their *primary biochemical effects.* These observations have led to the appreciation that while dysfunction within the monoaminergic neurotransmitter systems (which most current effective treatments target) is likely to play important roles in mediating some facets of the pathophysiology of mood and anxiety disorders, these dysfunctions likely represent the downstream effects of other, *more primary abnormalities* (Manji and Lenox 2000; Nestler 1998; Payne et al. 2002).

We should also keep in mind that a true understanding of the pathophysiology of these disorders must address their neurobiology at different physiological levels (i.e., molecular, cellular, systems, and behavioral) (Figure 5–1). Genetic abnormalities undoubtedly—at least in large part—underlie the neurobiology of the disorder at the molecular level, and this will become evident as we identify the susceptibility and protective genes for severe mood and anxiety disorders in the coming years. Once this has been accomplished, however, the even more difficult task of examining the impact of the faulty expression of these gene products (proteins) on integrated cell function must begin. It is at these different levels that investigators have recently identified several protein candidates using the psychopharmacological strategies that will be more fully elucidated in this chapter; the precise manner in which these candidate molecular targets may or may not relate to the faulty expression of susceptibility gene products is yet to be determined. The subsequent challenge for the basic and clinical neuroscientist will be the integration of these molecular and cellular changes to the systems and ultimately to the behavioral level wherein the clinical expression of mood and anxiety disorders becomes fully elaborated.

Despite formidable challenges, there has been considerable progress in our understanding of the underlying molecular and cellular basis of these disorders in recent years. In particular, recent evidence demonstrating that impairments of signaling pathways may play a role in the pathophysiology of mood disorders, and that antidepressants and mood stabilizers exert major effects on signaling pathways that regulate neuroplasticity and cell sur-

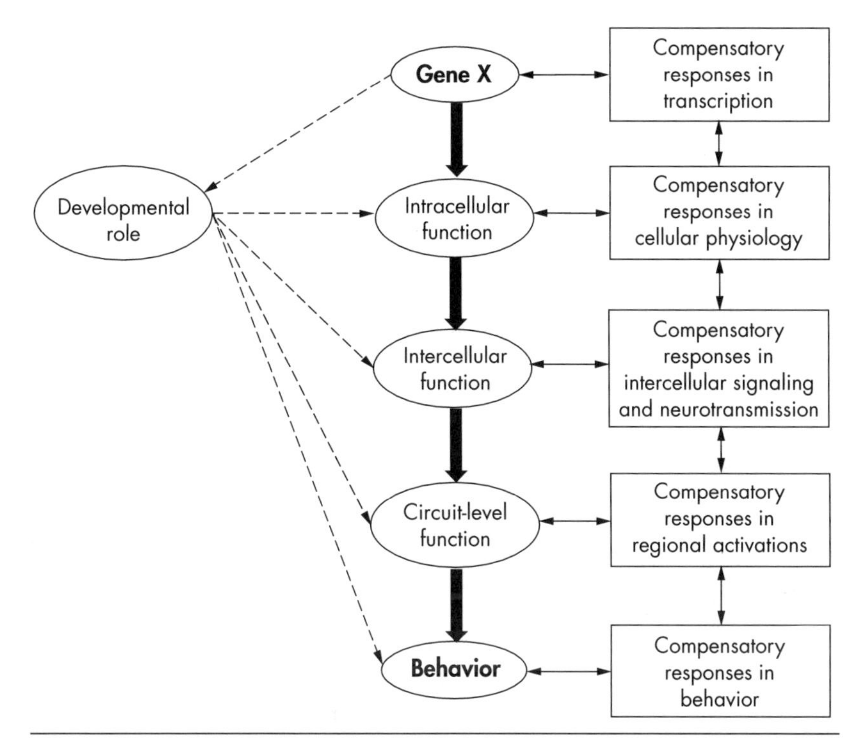


**Figure 5–1.** Interactions between genes, cellular and circuit physiology, and the environment.

vival, has generated considerable excitement among the clinical neuroscience community and is reshaping views about the neurobiological underpinnings of these disorders (Duman 2002; Gould et al. 2002; Manji and Chen 2002; Manji et al. 2001a; Nestler et al. 2002). In this chapter, we review these data and discuss their implications not only for changing existing conceptualizations regarding the pathophysiology of mood and anxiety disorders, but also for strategically developing improved therapeutics. It should be noted that abnormalities in multiple neurotransmitters, neuropeptides, and signaling pathways have been identified in mood and anxiety disorders in recent years. Space limitations preclude a discussion of all these findings here, and the interested reader is referred to several excellent reviews (Bezchlibnyk and Young 2002; Bremner and Charney 2002; Charney and Bremner 1999; Gould and Manji 2002a; Nemeroff 2002; Nemeroff and Owen 2002). In this chapter, we focus on those neurotransmitter/neuropeptide and signaling systems for which novel therapeutics for mood and anxiety disorders are already being explored.

## Signaling Networks: The Cellular Machinery Underlying Information Processing and Long-Term Neuroplastic Events

It is hardly surprising that abnormalities in multiple neurotransmitter systems and physiological processes have been found in these complex disorders. In this context, it is noteworthy that signal transduction pathways are in a pivotal position in the central nervous system (CNS) and are able to affect the functional balance between multiple neurotransmitter systems; they may therefore play a role in mediating the more "downstream" abnormalities that likely underlie the pathophysiology of affective disorders. Multicomponent, cellular signaling pathways interact at various levels, thereby forming complex signaling networks that allow the cell to receive, process, and respond to information (Bhalla and Iyengar 1999; Bourne and Nicoll 1993). These networks facilitate the integration of signals across multiple time scales, the generation of distinct outputs depending on input strength and duration, and the regulation of intricate feed-forward and feedback loops (Weng et al. 1999). These properties of signaling networks suggest they play critical roles in cellular memory; thus, cells with different "histories," and therefore expressing different repertoires of interacting signaling molecules, may respond quite differently to the same signal over time. It is not surprising that, given their widespread and crucial role in the integration, regulation, amplification, and fine-tuning of physiological processes, abnormalities in signaling pathways occur, which have now been identified in a variety of human diseases (Milligan and Wakelam 1992; Spiegel 1998; Weintraub 1995). Pertinent for the present discussion is the observation that a variety of human diseases arising from signaling abnormalities can manifest relatively circumscribed symptomatology, despite the widespread, often ubiquitous expression of the affected signaling proteins. This appears to be due to the stoichiometry of the signaling molecules in different tissues, the built-in redundancy, the ability of the tissue to compensate for the primary defect, and likely tissue-specific expression differences and genetic imprinting (see Weinstein 2001 for an excellent discussion of the latter phenomenon).

Although complex signaling networks are likely present in all eukaryotic cells and control various metabolic, humoral, and developmental functions, they may be especially important in the CNS, where they serve the critical roles of first amplifying and "weighting" numerous extracellularly generated neuronal signals and then transmitting these integrated signals to effectors, thereby forming the basis for a complex information-processing network (Bourne and Nicoll 1993; Manji 1992). These pathways are thus undoubtedly involved in regulating such diverse vegetative functions as mood, appetite, and wakefulness and are therefore likely to be involved in the pathophysiology of mood disorders (see Lenox et al. 2002) (Table 5–1). We now turn to a discussion of the direct and indirect evidence supporting a role for abnormalities in signaling pathways in the pathophysiology of mood disorders.

## Neurobiology of Mood Disorders

### *Protein Kinase C Signaling Pathway in the Pathophysiology and Treatment of Mood Disorders*

To date, there have been only a limited number of studies directly examining the protein kinase C (PKC) signaling cascade in mood disorders (Gould and Manji 2002a; Hahn and Friedman 1999) (Table 5–2). Although undoubtedly an oversimplification, particulate (membrane) PKC is sometimes viewed as the more active form of PKC, and thus an examination of the subcellular partitioning of this enzyme can be used as an index of the degree of activation. Friedman et al. (1993) investigated PKC activity and PKC translocation in response to serotonin in platelets obtained from subjects with bipolar disorder before and during lithium treatment. They reported that the ratio of PKC activities in the cell

**Table 5–1.** Putative role for signaling pathways in mood disorders

Signaling pathways
- Amplify, attenuate, and integrate multiple signals, the basis of intracellular circuits and cellular modules.
- Regulate multiple neurotransmitter and peptide systems, the basis of neuronal circuits and systems modules.
- Play critical role in cellular memory and long-term neuroplasticity.
- Are major targets for many hormones implicated in mood disorders, including gonadal steroids, thyroid hormones, and glucocorticoids.

Dynamic regulation of complex signaling networks form the basis for higher-order brain function, mood, and cognition.

Abnormalities *are* compatible with life—many human diseases arise from defects in signaling pathways.

Brain regional dysregulation and circumscribed symptoms are possible despite the relatively ubiquitous expression of signaling proteins.

Signaling proteins have been identified as targets for medications that are most effective in the treatment of mood disorders.

membrane compared with cell cytoplasm portions were elevated in the subjects with mania. In addition, serotonin-elicited platelet PKC translocation was enhanced in those subjects. In a postmortem brain study, increased PKC activity and translocation were found in bipolar disorder brains compared with control brains—effects that were accompanied by elevated levels of selected PKC isozymes (similar but distinct forms of the same enzyme) in the cortex of the bipolar disorder brains (see Wang and Friedman 1996).

Two more recent postmortem brain studies have utilized phorbol dibutyrate ([$^3$H]PDBu, a radioligand that binds to PKC) to investigate particulate and cytosolic PKC in postmortem brain samples obtained from depressed patients and/or individuals who committed suicide. One study found that number of [$^3$H]-PDBu binding sites was significantly decreased in both membrane and cytosolic fractions from Brodmann's areas 8 and 9 in teenage suicide subjects compared with matched controls (Pandey et al.

**Table 5–2.** Protein kinase C and the pathophysiology of bipolar affective disorder

| |
|---|
| Amphetamine produces an increase in PKC activity and GAP-43 phosphorylation (implicated in neurotransmitter release). |
| PKC inhibitors block the biochemical and behavioral responses to amphetamine and cocaine and also block cocaine-induced sensitization. |
| Membrane/cytosol PKC partitioning in platelets from manic subjects is increased and is normalized with lithium treatment. |
| PKC activity and translocation is increased in the brains of patients with bipolar disorder compared with those of controls. |
| Levels of RACK-1 (receptor for activated C kinase) are increased in the brains of bipolar disorder patients, compared with those of controls. |
| Lithium and valproate regulate PKC activity, PKC $\alpha$, PKC $\varepsilon$, and MARCKS. |
| PKC inhibitors may have efficacy in the treatment of acute mania (as suggested by preliminary data). |

*Note.* PKC = protein kinase C; MARCKS = myristoylated alanine-rich C kinase substrate.

1997), with the other reporting increased [$^3$H]PDBu binding in the soluble fraction (suggesting less in the active membrane fraction) in antidepressant-free suicide subjects compared with controls in the frontal cortex (Coull et al. 2000). The results of these two studies could *potentially* be interpreted as reflecting reduced PKC function, due to a reduction in either the absolute levels or the particulate/soluble fractions. However, considerable additional research is required to adequately justify such a conclusion.

## PKC in the Treatment of Bipolar Disorder

Evidence accumulating from various laboratories has clearly demonstrated that lithium, at therapeutically relevant concentrations, exerts major effects on the PKC signaling cascade (Gould et al. 2002) (Table 5–2). Currently available data suggest that acute lithium exposure facilitates a number of PKC-mediated responses, whereas longer-term exposure results in an attenuation of phorbol ester–mediated responses, which is accompanied by a downregulation of specific PKC isozymes (Manji and Lenox 1999). Studies in rodents—utilizing antibodies specific for different enzyme isozymes—have demonstrated that chronic (but not acute) lithium produces an isozyme-selective reduction in PKC $\alpha$ and $\varepsilon$ in frontal cortex and hippocampus, in the absence of significant alterations in the $\beta$, $\gamma$, $\delta$, or $\zeta$ isozymes (G. Chen et al. 2000; Manji et al. 1993; Manji and Lenox 1999). Furthermore, chronic lithium has been demonstrated to dramatically reduce the hippocampal levels of a major PKC substrate, MARCKS (myristoylated alanine-rich C kinase substrate), a protein that has been implicated in regulating long-term neuroplastic events (Lenox et al. 1992).

Although the effects of lithium on PKC isozymes and MARCKS are striking, a major problem inherent in neuropharmacological research is the difficulty in attributing *therapeutic relevance* to any observed biochemical finding. It is thus noteworthy that the structurally dissimilar antimanic agent valproate produces effects that are very similar to those of lithium on PKC $\alpha$ and $\varepsilon$ isozymes and MARCKS protein (G. Chen et al. 1994; Lenox and Hahn 2000; Manji and Chen 2000; Manji and Lenox 2000; Manji et al. 1999; Watson et al. 1998). Interestingly, lithium and valproate appear to bring about their effects on the PKC signaling pathway by distinct mechanisms (Lenox and Hahn 2000; Manji and Lenox 1999); these biochemical observations are consistent with the clinical observations that some patients show preferential response to one or other of the agents, and that one often observes additive therapeutic effects in patients when the two agents are coadministered.

In view of the pivotal role of the PKC signaling pathway in the regulation of neuronal excitability, neurotransmitter release, and long-term synaptic events (S. J. Chen et al. 1997; Conn and Sweatt 1994; Hahn and Friedman 1999), it was postulated that the attenuation of PKC activity may play a role in the antimanic effects of lithium and valproate. Recently, a pilot study found that tamoxifen, a nonsteroidal antiestrogen known to be a PKC inhibitor at higher concentrations (Baltuch et al. 1993), may indeed possess antimanic efficacy (Bebchuk et al. 2000). Clearly, these results have to be considered preliminary because of the small sample size thus far. However, in view of the preliminary data suggesting the involvement of the PKC signaling system in the pathophysiology of bipolar disorder, as discussed earlier, these results suggest that PKC inhibitors may be very useful agents in the treatment of mania. Thus, the results of larger, double-blind placebo-controlled studies of tamoxifen in the treatment of mania (currently under way) are eagerly awaited. We now turn to a discussion of an emerging field of research on mood disorders that has developed largely out of the use of recent molecular and cellular biological technologies: the study of signaling cascades regulating neuroplasticity and cellular resilience in the pathophysiology and treatment of recurrent mood disorders.

### *Impairments of Structural Plasticity and Cellular Resilience in Recurrent Mood Disorders*

Positron emission tomography (PET) imaging studies have revealed multiple abnormalities of regional cerebral blood flow (CBF) and glucose metabolism in limbic and prefrontal cortical (PFC) structures in mood disorders. These abnormalities implicate limbic-thalamic-cortical and limbic-cortical-striatal-pallidal-thalamic circuits, involving the amygdala, orbital and medial PFC, and anatomically related parts of the striatum and thalamus in the pathophysiology of mood disorders.

Interestingly, recent morphometric magnetic resonance imaging (MRI) and postmortem investigations have also demonstrated abnormalities of brain structure that persist independent of mood state and may contribute to the corresponding abnormalities of metabolic activity (discussed in Manji and Duman 2001; Manji et al. 2001a). Structural imaging studies have demonstrated reduced gray matter volumes in areas of the orbital and medial PFC, ventral striatum, and hippocampus, and enlargement of the third ventricle in samples of individuals with mood disorders relative to healthy control samples (reviewed in Beyer and Krishnan 2002; Drevets 2001; Strakowski et al. 2002). Also consistent is the presence of white matter hyperintensities in the brains of elderly depressed patients and in nonelderly bipolar patients when compared with control populations. White matter hyperintensities may have multiple causes, including cerebrovascular accidents, demyelination, loss of axons, dilated perivascular space, minute brain cysts, and necrosis, and recent studies have used diffusion tensor imaging (DTI) to study possible white matter tract disruption in mood disorders (Taylor et al. 2001). These studies support the contention that white matter hyperintensities reflect damage of the structure of brain tissue and, likely, disruption of the neuronal connectivity necessary for normal affective functioning. Although the cause of white matter hyperintensities in mood disorders is unknown, their presence—particularly in the brains of young bipolar patients—suggests importance in the pathophysiology of the disorder (Lenox et al. 2002; Stoll et al. 2000).

Complementary postmortem neuropathological studies have shown reductions in cortex volume, glial cell counts, and/or neuron size in the subgenual PFC, orbital cortex, dorsal anterolateral PFC, and amygdala (Table 5–3) (reviewed in Cotter et al. 2001; Manji and Duman 2001; Rajkowska 2002). It is not known whether these deficits constitute developmental abnormalities that may confer vulnerability to abnormal mood episodes, compensatory changes to other pathogenic processes, or the sequelae of recurrent affective episodes per se. Understanding these issues will partly depend on experiments that delineate the onset of such abnormalities within the illness course and determine whether they antedate depressive episodes in individuals at high familial risk for mood disorders. Nevertheless, the marked reduction in glial cells in these regions has been particularly intriguing in view of the growing appreciation of the critical roles that glia play in regulating synaptic glutamate concentrations and CNS energy homeostasis and in releasing trophic factors that participate in the development and maintenance of synaptic networks formed by neuronal and glial processes (Coyle and Schwarcz 2000; Haydon 2001; Ongur et al. 1998; Rajkowska 2000b; Rajkowska et al. 1999; Ullian et al. 2001). Abnormalities of glial function could thus prove integral to the impairments of structural plasticity and overall pathophysiology of mood disorders.

**Table 5–3.** Postmortem morphometric brain studies in mood disorders demonstrating cellular atrophy and/or loss

| |
|---|
| **Reduced volume/cortical thickness** |
| Cortical thickness in rostral orbitofrontal cortex in MDD (Rajkowska et al. 1999) |
| Laminar cortical thickness in layers III, V, and VI in subgenual anterior cingulate cortex (area 24) in bipolar disorder (Bouras et al. 2001) |
| Volume of subgenual prefrontal cortex in familial MDD and bipolar disorder (Ongur et al. 1998) |
| Volumes of nucleus accumbens (left) and basal ganglia (bilateral) in MDD and bipolar disorder (Baumann et al. 1999) |
| Parahippocampal cortex size (right) in suicide (Altshuler et al. 1990) |
| **Reduced neuronal size and/or density** |
| Pyramidal neuronal density and layers III and V in dorsolateral prefrontal cortex in bipolar disorder and MDD (Rajkowska 2000a) |
| Neuronal size in layer V (–14%) and VI (–18%) in prefrontal cortex (area 9) in bipolar disorder (Cotter et al. 2001) |
| Neuronal size in layer VI (–20%) in prefrontal cortex (area 9) in MDD (Cotter et al. 2001) |
| Neuronal density and size in layers II–IV in rostral orbitofrontal cortex, in layer V/VI in caudal orbitofrontal cortex, and in supra- and infragranular layers in dorsolateral prefrontal cortex, in MDD (Rajkowska et al. 1999) |
| Neuronal size in layer VI (–23%) in anterior cingulate cortex in MDD (Cotter et al. 2001) |
| Neuronal density in layers III, V, and VI in subgenual anterior cingulate cortex (area 24) in bipolar disorder (Bouras et al. 2001) |
| Layer-specific interneurons in anterior cingulate cortex in bipolar disorder and MDD (Vincent et al. 1997) |
| Nonpyramidal neuronal density in layer II (–27%) in anterior cingulate cortex in bipolar disorder (Benes et al. 2001) |
| Nonpyramidal neurons density in the CA2-region in bipolar disorder (Benes et al. 1998) |

**Table 5–3.** Postmortem morphometric brain studies in mood disorders demonstrating cellular atrophy and/or loss *(continued)*

| **Reduced glia** |
|---|
| Density/size of glia in dorsolateral prefrontal cortex and caudal oribitofrontal cortex in MDD and bipolar disorder—layer-specific (Miguel-Hidalgo and Rajkowska 2002; Rajkowska et al. 1999) |
| Glial cell density in sublayer IIIc (–19%) (and a trend to decrease in layer Va) in dorsolateral prefrontal cortex (area 9) in bipolar disorder (Selemon et al. 2002) |
| Glial number in subgenual prefrontal cortex in familial MDD (–24%) and bipolar disorder (–41%) (Ongur et al. 1998) |
| Glial cell density in layer V (–30%) in prefrontal cortex (area 9) in MDD (Cotter et al. 2001) |
| Glial cell density in layer VI (–22%) in anterior cingulate cortex in MDD (Cotter et al. 2001) |
| Glial cell counts, glial density, and glia-to-neuron ratios in amygdala in MDD (Bowley et al. 2002) |

*Note.* MDD = major depressive disorder.
*Source.* Modified from Manji and Duman 2001.

## *Stress and Glucocorticoids*

### Modulation of Neural Plasticity: Implications for Severe Mood Disorders

In developing hypotheses regarding the pathogenesis of these histopathological changes in mood disorders, the alterations in cellular morphology resulting from various stressors have been the focus of considerable recent research (D'Sa and Duman 2002) (Figure 5–2). Thus, although mood disorders undoubtedly have a strong genetic basis, considerable evidence has shown that severe stressors are associated with a substantial increase in risk for the onset of mood disorders in susceptible individuals.

Most studies of atrophy and survival of neurons in response to stress, as well as hormones of the hypothalamic-pituitary-adrenal (HPA) axis, have focused on the hippocampus. This focus is due, in part, to the well-defined and easily studied neuronal populations of this limbic brain region, which includes the dentate gyrus granule cell layer and the CA1 and CA3 pyramidal cell layers. Another major reason that the hippocampus has been the focus of stress research is that the highest levels of glucocorticoid receptors are expressed in this brain region (Lopez et al. 1998). However, it is clear that stress and glucocorticoids also influence the survival and atrophy of neurons in other brain regions that have not yet been studied in the same detail as the hippocampus (Lyons 2002).

One of the most consistent effects of stress on cellular morphology is atrophy of hippocampal neurons (McEwen 1999; Sapolsky 2000). The stress-induced atrophy of CA3 neurons (i.e., decreased number and length of the apical dendritic branches) occurs after 2–3 weeks of exposure to restraint stress or more long-term social stress and has been observed in rodents and tree shrews (see McEwen 1999; Sapolsky 2000). Atrophy of CA3 pyramidal neurons also occurs on exposure to high levels of glucocorticoids, suggesting that activation of the HPA axis likely plays a major role in mediating the stress-induced atrophy (see McEwen 1999; Sapolsky 1996, 2000). The hippocampus has a very high concentration of glutamate and expresses both type I and type II corticosteroid receptors, though type II receptors may be relatively scarce in the hippocampus of primates (Patel et al. 2000; Sanchez

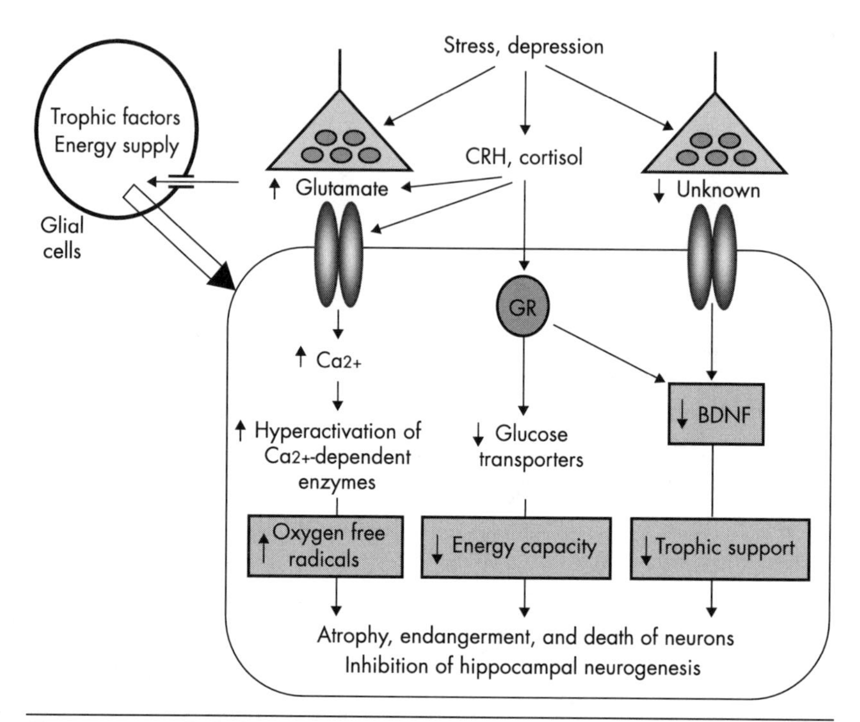


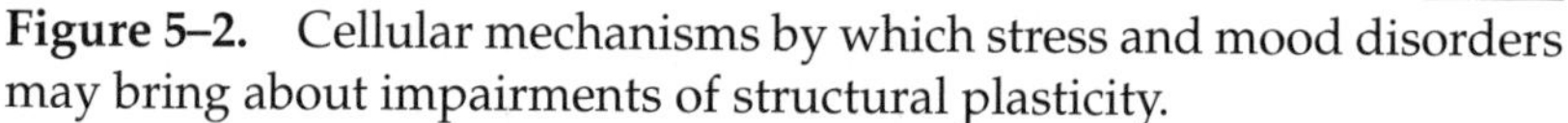

**Figure 5–2.** Cellular mechanisms by which stress and mood disorders may bring about impairments of structural plasticity.
This figure depicts the multiple mechanisms by which stress and potentially affective episodes may attenuate cellular resiliency, thereby resulting in atrophy, death, and endangerment of hippocampal neurons. The primary mechanisms appear to be i) excessive NMDA and non-NMDA glutamatergic throughput; ii) downregulation of cell surface glucose transporters, which are involved in bringing glucose into the cell. Reduced levels of glucose transporters thus reduce the neurons' energetic reservoirs, making them susceptible to energy failure when faced with excessive demands; and (iii) reduction in the levels of BDNF, which is essential for the neuron's normal trophic support and synaptic plasticity. The well-documented reduction in glial cells may contribute to impairments of neuronal structural plasticity by reducing the neuron's energy supply and through reduced glial-mediated clearing of excessive synaptic glutamate. BDNF = brain-derived neurotrophic factor; CRH = corticotropin-releasing hormone; GR = glucocorticoid receptor; NMDA = *N*-methyl-D-aspartate glutamate receptor.
*Source.* Modified from Manji and Duman 2001.

et al. 2000) and more abundant in cortical regions. Interestingly, mineralocorticoid, or type I, receptor (MR) activation in the hippocampus (CA1) is associated with reduced calcium currents, which has been postulated to attenuate the deleterious effects of excess excitatory amino acid (e.g., glutamate) release. A cautionary note in the interpretation of the clinical studies is suggested by recent primate studies which suggest that small hippocampi and ventral-medial prefrontal volumes may in large part also reflect an inherited characteristic of the brain, highlighting the difficulties with the attribution of causality in the cross-sectional human morphometric studies of the hippocampus (Lyons 2002; Lyons et al. 2001).

## Mechanisms Underlying Stress-Induced Morphometric Changes

Microdialysis studies have shown that stress increases extracellular levels of glutamate in hippocampus, and *N*-methyl-D-aspartate (NMDA) glutamate receptor antagonists attenuate stress-induced atrophy of CA3 pyramidal neurons (McEwen 1999; Sapolsky 2000). Overactivation of the glutamate ionotropic receptors is known to contribute to the neurotoxic effects of a variety of insults, including repeated seizures and ischemia (Figure 5–2). Activation of the HPA axis appears to play a critical role in mediating these effects, since stress-induced neuronal atrophy is prevented by adrenalectomy and duplicated by exposure to high concentrations of glucocorticoids (reviewed in McEwen 1999; Sapolsky 1996, 2000). Increasing recent data also suggest a critical role for corticotropin-releasing hormone (CRH) in long-term effects of early-life stress on hippocampal integrity and function (reviewed in Reul and Holsboer 2002; see "Neurobiology of Anxiety Disorders" later in this chapter for a more detailed description of the HPA axis and CRH).

Together, the abundant data for the critical roles of CRH and glucocorticoids are noteworthy with respect to the pathophysiology of mood disorders, since a significant percentage of patients with mood disorders display some form of HPA axis activation, and the subtypes of depression most frequently associated with HPA activation are those most likely to be associated with hippocampal volume reductions (Sapolsky 2000). Evidence in humans suggests that decreased corticosteroid receptor number (postmortem reduction of glucocorticoid receptor [GR] mRNA [Webster et al. 1999]) may be present in the hippocampus of individuals with bipolar and unipolar disorder, and some antidepressants (tricyclics), ECT, and mood stabilizers (lithium) may modulate GR number and/or function (reviewed in Holsboer 2000). Furthermore, transgenic mice with reduced GRs have HPA and cognitive disturbances that may parallel those seen in depression in humans and that normalize with antidepressant exposure. Finally, antisense oligonucleotides targeted to GRs reduce immobility on the forced swim test (a measure of depression-like behavior), as does the antiglucocorticoid drug mifepristone (RU 486) (Korte et al. 1996).

In addition to directly causing neuronal atrophy, stress and glucocorticoids also appear to reduce cellular resilience, thereby making certain neurons more vulnerable to other insults, such as ischemia, hypoglycemia, and excitatory amino acid toxicity (Sapolsky 2000). The potential functional significance of these effects is supported by the demonstration that overexpression of the glucose transporter blocks the neurotoxic effects of neuronal insults (Manji and Duman 2001; Sapolsky 2000). Such processes may conceivably also play a role in the relationship between mood disorders and cerebrovascular events, considering that individuals who develop their first depressive episode in late life have an increased likelihood of showing MRI evidence of cerebrovascular disease (Chemerinski and Robinson 2000; Drevets 2000; Kumar et al. 1997; Murray and Lopez 1997; Steffens and Krishnan 1998; Steffens et al. 1999). The precise mechanisms by which glucocorticoids exert these deleterious effects remain to be fully elucidated, but likely involve the inhibition of glucose transport (thereby diminishing capability of energy production and augmenting susceptibility to hypoglycemic conditions), and the aberrant, excessive facilitation of glutamatergic signaling (Sapolsky 2000). The reduction in the resilience of discrete brain regions, including hippocampus and potentially PFC, may also reflect the propensity for various stressors to decrease the expression of brain-derived neurotrophic factor (BDNF) in this region (Nibuya et al. 1999; Smith et al. 1995).

## Impairment of Hippocampal Neurogenesis

The demonstration that neurogenesis occurs in the adult human brain well into senescence has reinvigorated research into the cellular mechanisms by which the birth of new neurons is regulated in the mammalian brain (Eriksson et al. 1998). The localization of pluripotent progenitor cells and neurogenesis occurs in restricted brain regions. The greatest density of new cell birth is observed in the subventricular zone and the subgranular layer of the hippocampus. Cells born in the subventricular zone migrate largely to the olfactory bulb, and those in the subgranular zone into the granule cell layer of the hippocampus. The newly generated neurons send out axons and appear to make connections with surrounding neurons, indicating that they are capable of integrating into the appropriate neuronal circuitry in hippocampus and cerebral cortex. Neurogenesis in the hippocampus is increased by an enriched environment, exercise, and hippocampal-dependent learning (Gould et al. 2000; Kempermann 2002; van Praag et al. 1999). Upregulation of neurogenesis in response to these behavioral stimuli and the localization of this process to hippocampus have led to the proposal that new cell birth is involved in learning and memory (Gould et al. 2000).

Recent studies have shown that decreased neurogenesis occurs in response to both acute and chronic stress (Gould et al. 2000). Removal of adrenal steroids (i.e., adrenalectomy) increases neurogenesis, and treatment with high levels of glucocorticoids reproduces the downregulation of neurogenesis that occurs in response to stress. Aging also influences the rate of neurogenesis. Although neurogenesis continues into late life, the rate is significantly reduced (Cameron and McKay 1999). The decreased rate of cell birth could result from upregulation of the HPA axis and higher levels of adrenal steroids that occur in later life. Lowering glucocorticoid levels in aged animals restores neurogenesis to levels observed in younger animals, indicating that the population of progenitor cells remains stable but is inhibited by glucocorticoids (Cameron and McKay 1999). Interestingly, studies in glucocorticoid receptor knockout mice showed significant alterations in hippocampal neurogenesis (Gass et al. 2000). A reduction of granule cell neurogenesis (up to 65% of control levels) was found in MR–/– mice, whereas GR–/– mice did not show neurogenic disruption, eventually relating MRs in the pathogenesis of hippocampal changes observed in chronic stress and affective disorders (Gass et al. 2000). These observations raise the interesting possibility that CRH and GR antagonists, currently being developed for the treatment of mood and anxiety disorders, may have particular utility in the treatment of elderly depressed patients.

### *Neurotrophic Signaling Cascades*

Neurotrophins are a family of regulatory factors that mediate the differentiation and survival of neurons, as well as the modulation of synaptic transmission and synaptic plasticity (Patapoutian and Reichardt 2001; Poo 2001). The neurotrophin family now includes—among others—nerve growth factor (NGF), BDNF, and neurotrophin (NT) 3, NT4/5 and NT6 (Patapoutian and Reichardt 2001). These various proteins are related in terms of sequence homology and receptor specificity. They bind to and activate specific receptor tyrosine kinases belonging to the Trk family of receptors, including TrkA, TrkB, and TrkC, and a pan-neurotrophin receptor P75 (Patapoutian and Reichardt 2001; Poo 2001). Neurotrophins can be secreted constitutively or transiently, and often in an activity-dependent manner. Recent observations support a model wherein neurotrophins are secreted from the dendrite and generally act retrogradely at presynaptic terminals, where they act to induce long-lasting modifications (Poo 2001).

Within the neurotrophin family, BDNF is a potent physiological survival factor, which has also been implicated in a variety of pathophysiological conditions. The cellular actions of BDNF are mediated through two types of receptors: a high-affinity tyrosine receptor kinase (TrkB) and a low-affinity pan-neurotrophin receptor (P75). BDNF and other neurotrophic factors are necessary for the survival and function of neurons (Mamounas et al. 1995), implying that a sustained reduction of these factors could affect neuronal viability. However, what is sometimes less well appreciated is the fact that BDNF also has a number of much more *acute* effects on synaptic plasticity and neurotransmitter release, and it facilitates the release of glutamate, γ-aminobutyric acid (GABA), dopamine, and serotonin (Goggi et al. 2002; Matsumoto et al. 2001; Schinder et al. 2000).

As discussed already, BDNF is best known for its long-term neurotrophic and neuroprotective effects, which may be very important for its putative role in the pathophysiology and treatment of mood disorders. In this context, it is noteworthy that although endogenous neurotrophic factors have traditionally been viewed as increasing cell survival by providing necessary trophic support, it is now clear that their survival-promoting effects are mediated in large part by an inhibition of cell death cascades (Riccio et al. 1999). Increasing evidence suggests that neurotrophic factors inhibit cell death cascades by activating the mitogen activated protein (MAP) kinase signaling pathway and the phosphotidylinositol-3 kinase (PI-3K)/Akt pathway (Figure 5–3).

One important mechanism by which the MAP kinase signaling cascade inhibits cell death is by increasing the expression of the antiapoptotic protein bcl-2 (Bonni et al. 1999; Finkbeiner 2000). Bcl-2 is now recognized as a major neuroprotective protein, since overexpression of this protein protects neurons against diverse insults, including ischemia, MPTP (1-methyl-4-phenyl-1,2,3,6-tetrahydropyridine), β-amyloid, free radicals, excessive glutamate, and growth factor deprivation (Manji et al. 2001b). Accumulating data suggest that not only is bcl-2 neuropotective, it also exerts neurotrophic effects and promotes neurite sprouting, neurite outgrowth, and axonal regeneration (D. F. Chen et al. 1997; Chen and Tonegawa 1998; Chierzi et al. 1999; Holm et al. 2001; Oh et al. 1996). Moreover, a recent study demonstrated that severe stress exacerbates stroke outcome by suppressing bcl-2 expression (DeVries et al. 2001). In this study, the stressed mice expressed approximately 70% less bcl-2 mRNA than the unstressed mice after ischemia. Furthermore, stress greatly exacerbated infarct size in control mice but not in transgenic mice that constitutively express increased neuronal bcl-2. Finally, high corticosterone concentrations were significantly correlated with larger infarcts in wild-type mice but not in bcl-2–overexpressing transgenic mice. Thus, enhanced bcl-2 expression appears to be capable of offsetting the potentially deleterious consequences of stress-induced neuronal endangerment, and this suggests that pharmacologically induced upregulation of bcl-2 may have considerable utility in the treatment of a variety of disorders associated with endogenous or acquired impairments of cellular resilience (as discussed later in this chapter).

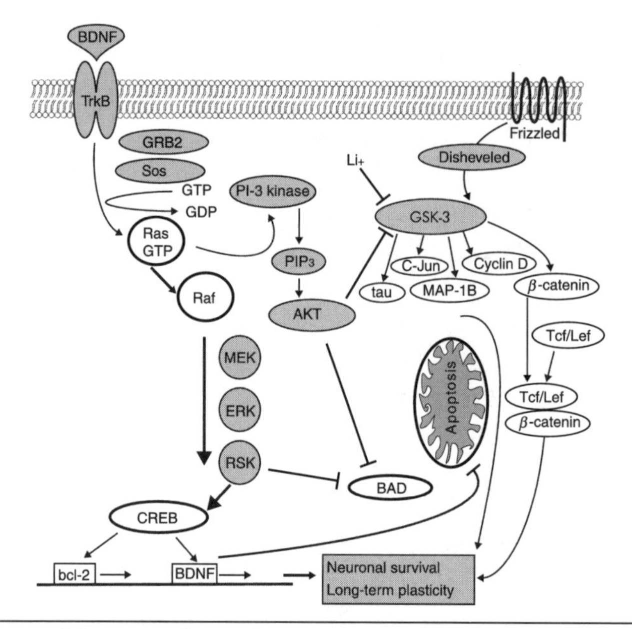


**Figure 5–3.** Neurotrophins, the ERK mitogen activated protein (MAP) kinase signaling cascade, and GSK-3.

Cell survival is dependent on neurotrophic factors, such as brain-derived neurotrophic factor (BDNF) and nerve growth factor, and the expression of these factors can be induced by synaptic activity. The influence of neurotrophic factors on cell survival is mediated by activation of the MAP kinase cascade. Activation of neurotrophic factor receptors, also referred to as Trks, results in activation of the MAP kinase cascade via several intermediate steps, including phosphorylation of the adaptor protein SHC and recruitment of the guanine nucleotide exchange factor Sos. This results in activation of the small guanosine triphosphate (GTP)–binding protein Ras, which leads to activation of a cascade of serine/threonine kinases. This includes Raf, MAP kinase kinase (MEK), and MAP kinase (also referred to as extracellular response kinase, or ERK). One target of the MAP kinase cascade is RSK, which influences cell survival in at least two ways. RSK phosphorylates and inactivates the pro-apoptotic factor BAD. RSK also phosphorylates cyclic adenosine monophosphate response element–binding protein (CREB) and thereby increases the expression of the anti-apoptotic factor bcl-2 and BDNF. These mechanisms underlie many of the long-term effects of neurotrophins, including neurite outgrowth, cytoskeletal remodeling, and cell survival. Recent evidence suggests that lithium and valproate (VPA) activate the ERK MAP kinase pathway. Lithium and VPA also appear to target GSK-3 and the Wnt signaling pathway. The function of GSK-3 in the Wnt signaling pathway—whereby Wnt glycoproteins interact with the frizzled family of receptors to stimulate the disheveled–mediated inactivation of GSK-3 and activation of the transcription factor β-catenin—is separate from the PI-3 kinase/AKT–mediated inactivation of GSK-3. Lithium is a direct inhibitor of GSK-3; both lithium and VPA increase β-catenin levels.

Overall, it is clear that the neurotrophic factor/MAP kinase/bcl-2 signaling cascade plays a critical role in cell survival in the CNS, and that there is a fine balance maintained between the levels and activities of cell survival and cell death factors; modest changes in this signaling cascade or in the levels of the bcl-2 family of proteins (potentially due to genetic, illness-, or insult-related factors) may therefore profoundly affect cellular viability. We now turn to a discussion of the growing body of data suggesting that neurotrophic signaling molecules play important roles in the treatment of mood disorders (Manji and Duman 2001).

## *Antidepressant Treatment and Neurotrophic Signaling Molecules*

### Influence on Cell Survival Pathways and Neurogenesis

Seminal studies from the Duman laboratory have investigated the possibility that factors involved in neuronal atrophy and survival could be the target of antidepressant treatments (D'Sa and Duman 2002; Duman et al. 1999) (Figure 5–4). These studies demonstrate that one pathway involved in cell survival and plasticity, the cAMP-CREB (cAMP response element binding protein) cascade, is upregulated by antidepressant treatment (Duman et al. 1999).

Upregulation of CREB and BDNF occurs in response to several different classes of antidepressant treatments, including norepinephrine and selective serotonin reuptake inhibitors (SSRIs), and electroconvulsive seizure, indicating that the cAMP-CREB cascade and BDNF are common postreceptor targets of these therapeutic agents (Nibuya et al. 1995, 1996). In addition, upregulation of CREB and BDNF is dependent on chronic treatment, consistent with the therapeutic action of antidepressants. A role for the cAMP-CREB cascade and BDNF in the actions of antidepressant treatment is also supported by studies demonstrating that genetic upregulation of these pathways increases performance in behavioral models of depression (Duman et al. 1999). Indirect (albeit very preliminary) human evidence comes from studies showing increased hippocampal BDNF expression in postmortem brain of subjects treated with antidepressants at the time of death compared with antidepressant-untreated subjects (B. Chen et al. 2001).

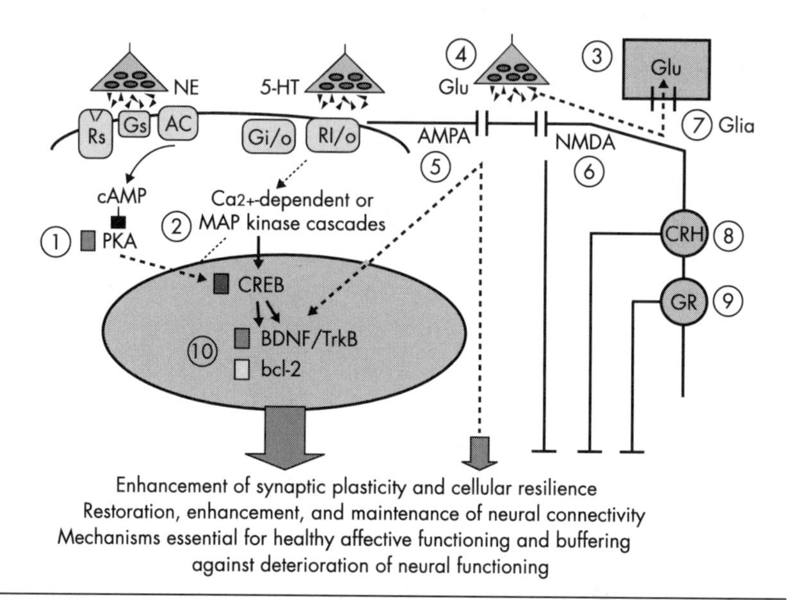


**Figure 5–4.** Plasticity regulators as targets for the development of novel agents for the treatment of depression.

This figure depicts the multiple targets by which neuroplasticity and cellular resilience can be increased in mood disorders. Genetic and neurodevelopmental factors, repeated affective episodes (and likely elevations of glucocorticoids), and illness progression may all contribute to the impairments of cellular resilience, volumetric reductions, and cell death/atrophy observed in mood disorders. Bcl-2 attenuates apoptosis by sequestering proforms of death-driving cysteine proteases (called *caspases*), by preventing the release of mitochondrial apoptogenic factors such as calcium, cytochrome c, and AIF (apoptosis-inducing factor) into the cytoplasm, and by enhancing mitochondrial calcium uptake. Antidepressants regulate the expression of BDNF (brain-derived neurotrophic factor), and its receptor TrkB. Both TrkA and TrkB use the PI-3 kinase/AKT and ERK MAP kinase pathways to bring about their neurotrophic effects. The ERK MAP kinase cascade also increases the expression of bcl-2 via its effects on CREB. (1) Phosphodiesterase inhibitors increase the levels of pCREB; (2) MAP kinase modulators increase expression of the major neurotrophic protein, bcl-2; (3) metabotropic glutamate 2/3 (mGluR2/3) agonists modulate the release of excessive levels of glutamate; (4) drugs such as riluzole and felbamate act on $Na^+$ channels to attenuate glutamate release; (5) AMPA potentiators upregulate the expression of BDNF; (6) NMDA antagonists like memantine enhance plasticity and cell survival; (7) novel drugs to enhance glial release of trophic factors and clear excessive glutamate may have utility for the treatment of depressive disorders; (8) CRH and (9) glucocorticoid antagonists attenuate the deleterious effects of hypercortisolemia, and CRH antagonists may exert other beneficial effects in the treatment of depression via non-HPA mechanisms; (10) agents that upregulate bcl-2 (e.g., pramipexole) would be postulated to have considerable utility in the treatment of depression and other stress-related disorders.

## Cellular/Neurotrophic Actions

There are several reports that support the hypothesis that antidepressants produce neurotrophic-like effects. One early study reported that antidepressant treatment induces a greater regeneration of catecholamine axon terminals in the cerebral cortex (Nakamura 1990), and the atypical antidepressant tianeptine was reported to block the stress-induced atrophy of CA3 pyramidal neurons (Watanabe et al. 1992). Most recently, Czeh et al. (2001) reported that stress-induced changes in brain structure and neurochemistry were also counteracted by treatment with tianeptine. In their study, male tree shrews subjected to a chronic psychosocial stress paradigm were found to have decreased *N*-acetylaspartate (NAA), a putative marker of neuronal viability (Tsai and Coyle 1995) measured in vivo by $^{1}$H–magnetic resonance spectroscopy (MRS), decreased granule cell proliferation in the dentate gyrus of the hippocampus, and a reduction in hippocampal volume as compared with nonstressed animals. These stress-induced effects were prevented or reversed in shrews treated concomitantly with tianeptine (Czeh et al. 2001); however, the generalizability of these intriguing findings to other, "more traditional" classes of antidepressants is unclear.

## Regulation of Hippocampal Neurogenesis

Recent studies have shown that chronic, but not acute, antidepressant treatment increases the neurogenesis of dentate gyrus granule cells (D'Sa and Duman 2002; Jacobs et al. 2000; Manev et al. 2001). These studies demonstrate that chronic administration of different classes of antidepressant treatments increases the proliferation and survival of new neurons, whereas cocaine, morphine, and haloperidol were without effect. Neurogenesis is increased by conditions that stimulate neuronal activity (e.g., enriched environment, learning, exercise), suggesting that this process is also positively regulated by, and may even be dependent on, neuronal plasticity (Kempermann 2002).

In view of the opposite effects of stress and antidepressants on hippocampal neurogenesis, it is quite plausible that alterations in hippocampal neurogenesis are fundamental to the pathophysiology and treatment of depression (D'Sa and Duman 2002; Jacobs et al. 2000; Kempermann 2002).

Although as yet there is no *direct* link between adult hippocampal neurogenesis and mood and emotion, the key role of the hippocampus in regulating diverse physiological and neurovegetative functions suggests that impairments of neurogenesis may also contribute to other facets of the clinical syndrome of depression (D'Sa and Duman 2002; Jacobs et al. 2000; Kempermann 2002). A critical next step will be the demonstration that inhibition of antidepressant-induced adult hippocampal neurogenesis results in an attenuation of the behavioral effects of antidepressants. In this context, it is noteworthy that genetic factors have also been clearly demonstrated to play a critical role in determining performance and the sensitivity to different types of antidepressant drugs in rodent models of depression (notably forced swim and tail suspension). Ongoing studies (G. Kempermann, personal communication, 2002) that are attempting to correlate antidepressant-induced changes in hippocampal neurogenesis in different mouse strains with performance in depression models should provide much needed information about the role of hippocampal neurogenesis in antidepressant drug action.

## *Mood Stabilizers and Neurotrophic Signaling Molecules*

### Robust Activation of Neurotrophic Signaling Cascades by Lithium and Valproate

As discussed in the preceding section, several endogenous growth factors—including nerve growth factor (NGF) and BDNF—exert many of their neurotrophic effects via the MAP kinase signaling cascade. In view of the important role of MAP kinases in mediating long-term neuroplastic events, it is noteworthy that lithium and valproate, at therapeutically relevant concentrations, have recently been demonstrated to robustly activate the extracellular signal-regulated kinase (ERK) MAP kinase cascade in human neuroblastoma SH-SY5Y cells and in critical limbic and limbic-related areas of rodent brain (Gould et al. 2002; Yuan et al. 2001). Consistent with an activation of neurotrophic signaling cascades, chronic treatment of rats with "therapeutic" doses of lithium and valproate produce a *doubling* of bcl-2 levels in frontal cortex—effects that were primarily due to a marked increase in the number of bcl-2 immunoreactive cells in layers II and III of frontal cortex (G. Chen et al. 2001; G. Chen et al. 1999b; Manji and Chen 2002). Interestingly, the importance of neurons in layers II–IV of the frontal cortex in mood disorders has recently been emphasized, since primate studies indicate that these areas are important for providing connections with other cortical regions and that they are targets for subcortical input (Rajkowska 2000a).

## Regulation of the GSK-3 Signaling Cascade by Lithium and Valproate

In addition to its robust effects on bcl-2, the effects of lithium and valproate on other signaling pathways and transcription factors may also contribute to their neuroprotective effects. Perhaps foremost among these is GSK-3 (Gould and Manji 2002b; Gould et al. 2002; Li et al. 2002) (Figure 5–3). Klein and Melton (1996) were the first to make the seminal observation that lithium, at therapeutically relevant concentrations, is an inhibitor of GSK-3. Valproic acid also has effects on GSK-3-mediated signaling events in cells (G. Chen et al. 1999a; Grimes and Jope 2001; Hall et al. 2002; Phiel et al. 2001). While lithium inhibits GSK-3 by direct competition with magnesium (Ryves and Harwood 2001), the precise mechanisms by which valproic acid exerts its action are still uncertain (Gould and Manji 2002b). GSK-3 not only controls developmental patterns in diverse organisms (including mammals) but also plays an important role in the mature CNS, by phosphorylating transcription factors and cytoskeletal proteins such as the Alzheimer's protein tau (a previous name for GSK-3 was tau kinase) (reviewed in Gould and Manji 2002b; Gould et al. 2002; Li et al. 2002; Phiel and Klein 2001). GSK-3 is rather unique among kinases in that it is constitutively active, and most intracellular signals to GSK-3 inactivate the enzyme. Signals deactivating GSK-3 arise from numerous growth factors, and it has been postulated that neurotrophic and neuroprotective effects are mediated, at least in part, by GSK-3 inhibition.

A rapidly increasing amount of evidence suggests that GSK-3 plays important roles in regulating neuroplasticity and cellular resilience, and there is thus considerable excitement about the development of GSK-3 inhibitors as novel therapeutic agents for bipolar disorder and classical neurodegenerative diseases. Although it was initially reported in 1993 that GSK-3 activity was required for β-amyloid induced neurotoxicity in primary hippocampal neurons (Takashima et al. 1993), these observations were not followed up until very recently. More recent studies have demonstrated that GSK-3 may regulate cell death beyond its role in β-amyloid induced toxicity. For example, GSK-3 overexpression induces apoptosis in cultured cells—a finding that is prevented by dominant negative GSK-3 mutants (Pap and Cooper 1998). Although the study of the effects of selective small molecule GSK-3 inhibitors is still in its infancy, the available data suggest that pharmacological inhibition of GSK-3β also exerts neuroprotective effects (Cross et al. 2001).

A second putative target pathway resultant from GSK-3 inhibition that may be of great relevance in the treatment of recurrent mood disorders is suggested by research exploring the underlying circadian cycle of *Drosophila* (discussed in detail in Gould and Manji 2002b; Lenox et al. 2002). The *Drosophila* orthologue of GSK-3 (SHAGGY) regulates circadian rhythms in this species. A decrease in SHAGGY activity results in an increase in circadian period length (Martinek et al. 2001), precisely the effect (increase in circadian period) that has been noted in numerous species, including *Drosophila,* after treatment with lithium (Klemfuss 1992). Abundant data suggest that abnormalities in biological rhythms may contribute to the pathophysiology of bipolar disorder; the ability of GSK-3 to regulate circadian rhythms provides additional support for the potential utility of GSK-3 inhibitors as novel therapeutic agents for bipolar disorder.

## *Evidence for Neuroprotective Effects of Mood Stabilizers*

Consistent with robust bcl-2 upregulation, ERK MAP kinase activation, and GSK-3 inhibition, lithium (and to a lesser extent valproate), at therapeutically relevant concentrations, has been shown to exert neuroprotective effects in a variety of preclinical paradigms (see Table 5–4). Thus, lithium has been demonstrated to protect against the deleterious effects of glutamate, NMDA receptor activation, aging, serum/nerve growth factor deprivation, ouabain, thapsigargin (which mobilizes intracellular $MPP^+$ [1-methyl-4-phenylpyridinium]), $Ca^{2+}$ and β-amyloid in vitro (Manji and Chen 2002). More importantly, lithium's neurotrophic and cytoprotective effects have also been demonstrated in rodent brain in vivo. Thus, lithium treatment has been shown to attenuate the biochemical deficits produced by kainic acid infusion, quinolinic acid infusion, ibotenic acid infusion, and forebrain cholinergic system lesions; to exert dramatic protective effects against middle cerebral artery occlusion (Manji and Chen 2002); and to enhance hippocampal neurogenesis in the adult rodent hippocampus (G. Chen et al. 2000).

## Human Evidence for the Neurotrophic Effects of Lithium

While the body of preclinical data demonstrating neurotrophic and neuroprotective effects of mood stabilizers is striking, considerable caution must clearly be exercised in extrapolating to the clinical situation with humans. In view of lithium's robust effects on the levels of the cytoprotective protein bcl-2 in the frontal cortex, Drevets and associates (1997) reanalyzed older data demonstrating approximately 40% reductions in subgenual PFC volumes in familial mood disorder subjects. Consistent with neurotrophic/neuroprotective effects of lithium, they found that the patients treated with chronic lithium or valproate exhibited subgenual PFC volumes, which were significantly higher than the volumes in non lithium- or valproate-treated patients, and not significantly different from those in the control subjects (W. C. Drevets, personal communication, 2000). In a more recent study, Drevets and colleagues investigated glial cell densities in mood disorder patients. Although the sample sizes were quite small, they made the intriguing observation that patients with unipolar mood disorders exhibited reduced glial cell densities, whereas only the patients with bipolar disorders who were not taking chronic lithium or valproate exhibited similar reductions (Bowley et al. 2002).

**Table 5–4.** Neurotrophic and neuroprotective effects of lithium

| |
|---|
| **Rodent and human studies** |
| Protects cultured cells of rodent and human neuronal origin in vitro from |
| Glutamate, NMDA |
| High concentrations of calcium |
| $MPP^+$ |
| β-Amyloid |
| Aging-induced cell death |
| HIV regulatory protein, Tat |
| Glucose deprivation |
| Growth factor or serum deprivation |
| Toxic concentrations of anticonvulsants |
| Platelet activating factor |
| Aluminum toxicity |
| Low $K^+$ |
| C2-ceramide |
| Ouabain |
| GSK-3β + staurosporine/heat shock |
| **Rodent studies** |
| Enhances hippocampal neurogenesis in adult mice |
| Protects rodent brain in vivo from |
| Cholinergic lesions |
| Radiation injury |
| Middle cerebral artery occlusion (stroke model) |
| Quinolinic acid (Huntington's model) |
| **Human studies** |
| No subgenual PFC gray matter volume reductions in cross-sectional MRI studies |
| No reductions in amygdala glial density in postmortem cell-counting |
| Increased total gray matter volumes on MRI of BD patients compared with untreated BD patients in cross-sectional studies |
| Increased NAA (marker of neuronal viability) levels in BD patients in longitudinal studies |
| Increased gray matter volumes in BD patients in longitudinal studies |

*Note.* BD = bipolar disorder; MRI = magnetic resonance imaging; NAA = *N*-acetylaspartate; NMDA = *N*-methyl-D-aspartate; PFC = prefrontal cortical.

Although the results of the aforementioned studies suggest that mood stabilizers may have provided neuroprotective effects during naturalistic use, considerable caution is warranted in view of the small sample sizes and cross-sectional nature of the studies. To investigate the potential neurotrophic effects of lithium in humans more directly, a longitudinal clinical study was recently undertaken, using proton MRS to quantitate NAA (a putative marker of neuronal viability [Tsai and Coyle 1995]) levels. Four weeks of lithium treatment produced a significant increase in NAA levels, effects that were localized almost exclusively to gray matter (Moore et al. 2000a). These findings provide intriguing indirect support for the contention that chronic lithium increases neuronal viability/function in the human brain. Furthermore, an approximately 0.97 correlation between lithium-induced NAA increases and regional voxel gray matter content was observed, thereby providing evidence for co-localization with the regional specific bcl-2 increases observed (e.g., gray vs. white matter) in the rodent brain cortices. These results suggest that chronic lithium may exert not only robust neuroprotective effects (as has been demonstrated in a variety of preclinical paradigms) but also neurotrophic effects in humans (Table 5–4).

A follow-up volumetric MRI study has demonstrated that 4 weeks of lithium treatment also significantly increased *total* gray matter content in the human brain (Moore et al. 2000b), suggesting an increase in the volume of the neuropil (the mosslike layer comprising axonal and dendritic fibers that occupies much of the cortex gray matter volume). A finer-grained, subregional analysis of these brain imaging data is ongoing and clearly shows that lithium produces a regionally selective increase in gray matter, with prominent effects being observed in hippocampus and caudate (Moore et al. 2001). Furthermore, no changes in overall gray matter volume are observed in healthy volunteers treated chronically with lithium, suggesting that lithium is truly producing a reversal of illness-related atrophy, rather than nonspecific gray matter increases.

Overall, there is now an impressive body of data suggesting that recurrent mood disorders are associated with impairments of neuroplasticity and cellular resilience, effects that may arise (at least in part) from abnormalities of CRH/glucocorticoid, glutamatergic, and neurotrophic signaling systems. We now turn to a discussion of strategies being utilized to investigate the putative efficacy of plasticity enhancers as novel antidepressant agents.

## The Glutamatergic System as a Target for Novel Antidepressant Treatments

A growing body of preclinical and clinical research suggests that the NMDA class of glutamate receptors may be involved in the pathophysiology of mood disorders and the mechanism of action of antidepressants (Table 5–5) (reviewed in Krystal et al. 2002; Skolnick 1999; Zarate et al., in press). Thus, NMDA antagonists have demonstrated antidepressant effects in animal models of depression, and antidepressants have been shown to affect NMDA receptor subunit mRNA levels, NMDA radioligand binding, and NMDA function. Recently, a placebo-controlled, double-blind trial found that depressed subjects showed significant improvement in depressive symptoms within a short period of time (72 hours) after taking ketamine (an NMDA antagonist) but not placebo (Berman et al. 2000). In addition, the ketamine-induced mood improvement persisted for 1–2 weeks after the infusion. Similarly, another antiglutamatergic agent, lamotrigine, has been found to be effective in acute bipolar depression (Calabrese et al. 1999).

Several novel glutamatergic strategies are currently being investigated in the treatment of depression; these are briefly discussed here (for a complete review see Zarate et al., in press). Riluzole, a neuroprotective agent with anticonvulsant properties, is currently approved by the Food and Drug Administration (FDA) for the treatment of amytrophic lateral sclerosis (ALS). It acts by inhibition of glutamate release (by blocking presynaptic calcium and sodium ion channels); by inhibition of the voltage-dependent sodium channels in mammalian CNS neurons; and by inhibition of postsynaptic actions by indirect mechanisms. It has been shown to have neuroprotective properties in animal models of Parkinson's disease, dementia, ischemia, and traumatic CNS injury. A pilot study in progess is investigating the putative antidepressant effects of riluzole in unipolar and bipolar depression (Zarate et al., in press).

**Table 5–5.** Direct and indirect evidence supporting a role for the glutamatergic system in the pathophysiology and treatment of mood disorders

| |
|---|
| **CSF, plasma, and platelet studies** |
| CSF glutamine levels elevated in medication-free depressed patients vs. control patients (2 with BD, 16 with MDD) and correlated with CSF magnesium level (Levine et al. 2000) |
| Higher glutamine plasma levels in 59 depressive patients (MDD, BD) vs. controls (Mathis et al. 1988) |
| Increased glutamate plasma levels and decreased platelet levels in medication-free depressed patients (4 with MDD, 11 with BD) vs. controls (Altamura et al. 1993) |
| **MRS studies** |
| Increased glutamate/glutamine ratio in frontal lobe and basal ganglia in children with BD (Castillo et al. 2000) |
| Decreased glutamate levels in anterior cingulate cortex of depressed patients vs. controls (1 with BD and 18 with MDD; 7 patients were medication free and 12 were receiving antidepressant treatment) (Auer et al. 2000) |
| **Postmortem brain studies** |
| Decreased neuronal EAAT3 and EAAT4 mRNA expression in striatum in BD (McCullumsmith and Meador-Woodruff 2002) |
| Higher NR2D (a subunit of NMDA receptor) mRNA in striatum in BD ($n = 15$) vs. MDD ($n = 15$) (Meador-Woodruff et al. 2001) |
| Lower gluR1 (a subunit of AMPA receptor) mRNA in BD ($n = 15$) vs. controls ($n = 15$) (Meador-Woodruff et al. 2001) |
| [$^3$H]AMPA binding was higher in BD vs. MDD (Meador-Woodruff et al. 2001) |
| Reduced glutamate decarboxylase immunoreactive marked terminals in anterior cingulate cortex (most in layers II–III) (Benes et al. 2000) |

**Table 5–5.** Direct and indirect evidence supporting a role for the glutamatergic system in the pathophysiology and treatment of mood disorders *(continued)*

| **Treatment-related findings** |
| --- |
| *Preclinical studies* |
| Tricarboxylic acid cycle and ATP are reduced with valproate in mice (Johannessen et al. 2001). |
| Chronic lithium up-regulates and stabilizes glutamate uptake by presynaptic nerve endings in mouse cerebral cortex (Dixon and Hokin 1998). |
| Chronic antidepressants regulate NMDA receptor mRNA and binding (Boyer et al. 1998). |
| Imipramine and phenelzine decrease potasium-stimulated glutamate outflow in rat prefrontal cortex and not in striatum (Michael-Titus et al. 2000). |
| AMPA receptor potentiator LY392098 (a biarylpropylsulfonamide) produces an antidepressant-like effect in rats and mice (Li et al. 2001). |
| *Clinical studies* |
| Lamotrigine is effective in treatment-resistant BD (Sporn and Sachs 1997). |
| Lamotrigine has significant antidepressant efficacy in 195 depressed patients with BD I (double-blind, placebo-controlled study) (Calabrese et al. 1999). |
| Lamotrigine is effective as an antimanic agent (Berk 1999). |
| Lamotrigine showed a response rate of 52%–63% in depressed, manic, mixed, and rapid-cycling BD patients (Hurley 2002). |

**Table 5–5.** Direct and indirect evidence supporting a role for the glutamatergic system in the pathophysiology and treatment of mood disorders *(continued)*

| |
|---|
| **Treatment-related findings** *(continued)* |
| *Clinical studies* (continued) |
| Density of flumazenil binding site $GABA_A$ R ($\gamma_2$ receptor subunit) is increased in Brodmann's area 9 (Dean et al. 2001). |
| Ketamine improves depressive symptoms in depressed patients (8 with MDD, 1 with BD) lasting longer (3 days) than euphoric effects (hours) (double-blind, placebo-controlled study) (Berman et al. 2000). |

*Note.* BD = bipolar disorder; CSF = cerebrospinal fluid; $GABA_A$ = γ-aminobutyric acid receptor A; MDD = major depressive disorder; MRS = magnetic resonance spectroscopy; NMDA = *N*-methyl-D-aspartate.

Felbamate is a broad-spectrum anticonvulsant chemically related to meprobamate. It acts as a noncompetitive NMDA antagonist (NR1-2B), through inhibition of AMPA/kainate and dihydropyridine-sensitive calcium channels and enhancement of $GABA_A$ activity. It has been shown to have neuroprotective properties in hypoxia-ischemia models, in hippocampal traumatic injury models, and in culture from excitotoxic insult. Studies are currently under way that are investigating the potential antidepressant efficacy of felbamate in patients with treatment-refractory (due to its rare but serious side effects of aplastic anemia and hepatotoxicity) depression (Zarate et al., in press).

Memantine has anticonvulsant and neuroprotective properties and is approved in Germany for use in treating mild and moderate cerebral performance disorders with concentration and memory disorders, loss of interest and drive, premature fatigue, and depressive mood (dementia syndrome); cerebral and spinal spasticity; and Parkinson's disease. It acts as a modulator of glutamatergic neurotransmission: in the state of a reduced glutamate release, after degeneration of neurons, memantine results in an improvement in signal transmission and activation of neurons; in the state of a massive glutamate release (e.g., ischemia), memantine blocks NMDA receptors that mediate the excitotoxic action of glutamate on neurons. Memantine is well tolerated; side effects include dizziness, internal and motoric restlessness and agitation, fatigue, congestion in the head, and nausea. Only a few isolated cases of psychosis and cognitive deficits have been reported with its use. A large double-blind, placebo-controlled trial examining the efficacy of memantine in the treatment of depression is under way (Zarate et al., in press).

### *The HPA Axis as a Target for the Development of Novel Therapeutics*

As discussed earlier in this chapter, there is a growing appreciation that abnormalities of the HPA axis may play a role in the mediating the phenotypic expression of certain depressive states (Gold and Chrousos 2002), and not surprisingly, there is increasing interest in targeting this system for the development of novel therapeutics. Novel therapeutic targets for the treatment of dysfunction of the HPA system include dehydroepiandrosterone (DHEA), steroid synthesis inhibitors, CRH antagonists, and glucocorticoid antagonists (McQuade and Young 2000). Treatment with glucocorticoid synthesis inhibitors (e.g., ketoconazole) has been observed to rapidly ameliorate depressive symptoms in patients with treatment-resistant depression (Reus and Wolkowitz 2001). Clinical studies are also investigating the putative efficacy of the GR antagonist mifepristone for the treatment of depression. Mifepristone is a potent antagonist of GRs (but not MRs) as well as progesterone receptors. Empirical evidence of its efficacy in rapid treatment of depression in patients with Cushing's disorder inspired further study of its use for primary affective disorders. There are a number of mechanistic considerations suggesting that a more extensive investigation of the potential efficacy of mifepristone to protect against the negative effects of excessive GR signaling in depression is warranted (see Gold and Charney, in press). Among them, mifepristone may antagonize the effects of excessive cortisol on PFC and hippocampal structure and function (which leads to disinhibition of the HPA axis), thereby interrupting a maladaptive feed-forward mechanism in severely depressed patients. Additionally, the amygdala inhibits the PFC and activates both the HPA axis and brain-stem noradrenergic nuclei (including the locus coeruleus); thus, mifepristone, by blocking amygdala GRs, may attenuate the deleterious effects of amygdala hyperactivity on the PFC, hippocampus, and brainstem sites. Preliminary data indicate that mifepristone may be rapidly and robustly effective in the treatment of psychotic depression, the subtype of depression most clearly associated with HPA axis hyperactivity (Belanoff et al. 2001). Additional preliminary data from a cross-over bipolar depression study also suggested suggested functional improvement after only brief therapy (Young 2002). A larger, double-blind, placebo-controlled study of mifepristone in inpatients with bipolar depression is being undertaken (Manji et al., in press).

Further preclinical study investigating small-molecule CRH antagonists for the treatment of depression is also under way. The orally active CRH antagonist antalarmin significantly reduces fear and anxiety responses in nonhuman primates (Habib et al.

2000) and may be a candidate for further development. Since standard antidepressants appear to take several weeks to suppress CRH expression and release, the blockade of CRH receptors has been proposed as a rapid antidepressant strategy (Zobel et al. 2000). Several pharmaceutical companies have developed substituted pyrimidine small-molecule CRH antagonists (for review, see Owens and Nemeroff 1999). In an open-label dose escalation study, 24 subjects with major depression received either a high dose or a low dose of the CRH antagonist R121919; in the high-dose group ($n = 10$), 8 of 10 subjects analyzed were responders and 6 of 10 were remitters, compared with 5 of 10 responders and 3 of 10 remitters in the low-dose group. Patients had an abrupt worsening of mood symptoms when the antagonist was discontinued, suggesting that response was pharmacologically contingent. R121919, although generally well tolerated, was dropped from clinical development because of hepatotoxicity; however, it provides proof of concept in a preliminary fashion for this strategy.

### *Strategies to Potentiate the CREB/BDNF/bcl-2 Cascade for the Treatment of Depressive Disorders*

As discussed earlier, a growing body of data suggesting that agents which modulate the CREB/BDNF/bcl-2 cascade, may have utility for the treatment of mood disorders (D'Sa and Duman 2002; Nestler et al. 2002; Quiroz and Manji 2002). One approach is to use an inhibitor of phosphodiesterase (PDE), the enzyme responsible for the breakdown of cAMP. In preclinical studies in rats, Takahashi et al. (1999) demonstrated that chronic antidepressant administration increases the expression of the cAMP-specific PDE4A and PDE4B isoforms; these effects likely represent a compensatory "counter-regulatory" response to the chronic antidepressants (Nibuya et al. 1996). These data suggest that PDE4A and PDE4B may be relevant targets for development of agents that possess antidepressant effects either as monotherapy or in combination with agents that increase intrasynaptic monoamine levels, due to the possible synergism of effects on the cAMP cascade. Additional preclinical studies conducted to date support the potential utility of PDE4 inhibitors in the treatment of depressive disorders. Thus, long-term administration of PDE4 inhibitors increases the expression of CREB and BDNF in the hippocampus of rats (Asanuma et al. 1996; Fujimaki et al. 2000), and PDE inhibitors have antidepressant-like effects in behavioral models (Griebel et al. 1991; O'Donnell 1993; Wachtel and Schneider 1986).

In the 1980s and early 1990s, a number of open (Zeller et al. 1984) and controlled (Bertolino et al. 1988; Bobon et al. 1988; Fleischhacker et al. 1992; Hebenstreit et al. 1989) clinical trials demonstrated that rolipram, a specific inhibitor of the high-affinity cAMP PDE4, may have antidepressant efficacy in depressed patients. In addition, there is evidence that rolipram may have a faster onset of response compared with the standard antidepressants. In the clinical study by Horowski and Sastre-Y-Hernandez (1985), a rapid antidepressant effect to rolipram was reported to occur within the first 12 hours to 10 days, and another study (Zeller et al. 1984) reported an antidepressant effect only after 2 to 4 days of treatment. However, in a randomized, double-blind comparative trial of rolipram versus imipramine, Hebenstreit et al. (1989) found that imipramine was more effective than rolipram, though both were beneficial. Further, in another study, Scott and colleagues (1991) concluded that amitriptyline was more effective than rolipram in the treatment of depressed inpatients. Though the overall literature on the use of rolipram is suggestive that PDE inhibitors may have antidepressant efficacy and may have a faster onset of action, the potential use of rolipram for depression was limited because of often intolerable side effects such as nausea and emesis.

Recent studies demonstrating that PDE4 is expressed in inflammatory cells such as eosinophils, and that inhibition of PDE4 downregulates the inflammatory response, has generated renewed excitement about the possible utility of this class of agents in the treatment of diseases such as asthma, chronic obstructive pulmonary disease (COPD), rheumatoid arthritis, Crohn's disease, and multiple sclerosis (MS) (Dyke and Montana 2002; Huang et al. 2001). Second-generation compounds with markedly improved tolerability are rapidly being developed (Dyke and Montana 2002;

Huang et al. 2001), and it is anticipated that the availability of CNS-penetrant PDE4 inhibitors—perhaps initially developed for these other diseases—may lead to the development of a novel class of antidepressants.

## Neurobiology of Anxiety Disorders

### *Animal Models of Fear and Anxiety*

Animal models of anxiety have used primarily the rat, the mouse, and, to a lesser extent, nonhuman primates. It is not particularly difficult to evoke or measure anxiety in these species. However, difficulties arise when one attempts to define exactly how a stimulus and resultant behavioral response are related to human behavior: When a mouse exhibits freezing behavior to an unfamiliar and threatening cue, what is the human "equivalent"? Or, similarly, what stimulus could one present to a rat to best model the anxiety-inducing experience of public speaking? The unique set of symptoms and physiological markers associated with each anxiety disorder makes the development of truly specific models difficult. There is also the extremely important distinction between models that simply measure the anxious response of normal animals exposed to pharmacological or other manipulation, as might be used to screen drugs or genes, and those that attempt to create a pathological model of abnormal behavior.

#### Issues of Validity of Models

There are three generally accepted criteria for validating animals models for human psychiatric disorders: face validity, construct validity, and predictive validity. *Face validity* refers to the outward appearance of the model—namely, does the animal's behavior adequately reflect the human behavior being modeled? In this dimension, anxiety models have a clear advantage over other psychiatric models. It is usually quite apparent if an animal is frightened; it is a much more difficult to assess whether an animal is displaying psychotic-like or depressive-like behavior, for example. *Construct validity* refers to the fundamental causality and etiology of the behavior. Again, animal models of anxiety perform relatively well in this regard; the purpose and underlying anatomical and neurochemical substrates of anxiety are well conserved. However, although the general modeling of fear and anxiety has both face and construct validity, models of specific disorders are much more meager in these respects.

*Predictive validity* is the ability of a model to predict the effect that pharmacological or other manipulations will have on the condition being modeled. This criterion can present a real difficulty, in that drug development is often dictated by animal models. For example, if a given model detects only a subset of efficacious compounds (i.e., those belonging to a specific chemical class), then useful candidates will be discarded long before clinical trials, and the flaw in the model's predictive validity will not be discovered. Thus, the possibility that a model will yield false negative results is difficult to rule out.

Currently, there are a wide variety of methods employed to measure animal anxiety. In rodents, the most commonly used measures employ these animals' natural fear of brightly lit, open spaces. Measures may include exploratory behavior and social behaviors (in the social interaction test), with the expectation that such behaviors will decrease as anxiety increases. Advantages of these methods, including light-dark exploration and open-field tests, lie primarily in their ease of use. While these models have held great predictive validity for benzodiazepines and barbiturates, they have been only marginally successful at predicting the effects of serotonergic and noradrenergic anxiolytics. These methods are also currently enjoying great popularity in the behavioral testing of transgenic and knockout mice.

In operant conflict models, animals are trained to respond for a reward, such as food. The reward is subsequently paired with a punishment (e.g., foot shock), causing a reduction in response frequency. The degree to which the animal ceases to respond is thought to reflect its anxiety about receiving punishment. The predictive validity of this method is somewhat better than the elevated plus maze and social interaction measures, in that chronic administration of antidepressants or serotonin$_{1A}$ (5-HT$_{1A}$) partial agonists results in an anxiolytic effect, at least in rats. Furthermore, the requirement for chronic treatment with these drugs appears to be preserved in this model. However, the long training periods and specialized equipment required for this technique make it impractical for many laboratories.

Ultrasonic vocalizations emitted by rat pups (under the age of 14 days) when they are isolated from their mother are thought to reflect anxiety. This measure has proven sensitive to both anxiolytic and anxiogenic manipulation of GABA neurotransmission. However, the early developmental window used is problematic, in that chronic drug administration probably results in a variety of compensatory changes not seen in adulthood and may alter development of relevant brain systems. Indeed, in contrast to the clinical situation, the antidepressant clomipramine has acute, but not chronic, anxiolytic efficacy in this model.

Nonhuman primates offer another, although clearly much more difficult, avenue in investigating fear and anxiety (Barros and Tomaz 2002). The interspecies variation in neurobiology and behavior between nonhuman primates and humans is substantially reduced compared with rodents. A variety of measures have been developed, with use of stressors such as social isolation, conspecific intruder, extraspecific (human or predator) threat, and pharmacological anxiogenesis. These models have demonstrated validity and may be especially valuable in the later stages of preclinical drug development.

In conclusion, animal models allow us to study the biology of anxiety behavior and the manipulation of that behavior through environmental, pharmacological, and genetic mechanisms. However, the ability of these models to adequately replicate the character and diversity of human anxiety disorders is far from clear. Given the rapid advancement of techniques for probing human brain function in vivo, we will increasingly be able to discern disorder-specific biology, and thereby develop more valid animal models.

## Genetic Studies of Anxiety in Mice

The ability to manipulate the genome—either through adding a gene (transgenic) or removing a gene (knockout)—of mice has allowed an entirely novel and remarkable type of study: to relate a particular gene directly to a behavior. From these genomic manipulation strategies, a large number of genes have been implicated, to some degree, in the pathophysiology of anxiety (summarized in Appendix 1 at end of this chapter). In principle, the knowledge that can be gained by such an approach is impressive. Given sufficient knowledge about both the gene and the behavior(s) in question, it is possible to simultaneously move "bottom up" and "top down"; that is, one may proceed upward from gene function to the cell biology/physiology level, but also proceed downward from the behavioral to systems/circuit–level neuroscience (Figure 5–1). In this way, one may begin to attack from both sides the "black box" that is the complexity between genes and behavior. However, the fact remains that genes and behaviors cannot be associated on a simplistic, one-to-one basis; the true relationship between a gene and a behavior is probably more akin to chaos theory's "sensitive dependence on initial conditions." For example, there is presumably no gene for "language"; there are a number of genes that pattern the embryonic brain in such a way as to facilitate, and allow the physiological processes necessary for, language acquisition. In a similar regard, no gene has been found to singularly code for a human psychiatric condition. Indeed, it is unlikely that any single gene does code for a psychiatric condition per se, but rather that susceptibility genes interacting with developmental factors, both day-to-day and profound environmental events, epigenetic DNA modifications, and possibly entirely stochastic mechanisms eventually lead to the development of normal and abnormal human behaviors.

While many difficulties exist in the interpretation of knockout/transgenic studies (Bilbo and Nelson 2001), one of the most troubling is the relative role of immediate versus developmental and adaptive effects. Many knockout studies have been interpreted in much the same way as a pharmacological manipulation, assuming that the only difference between the wild-type and experimental animal is the presence or absence of a specific protein. However, the actual effect of a knockout can be far-reaching and highly complex. Given the universality of homeostatic regulation in biology, from the intracellular to the organismic level, it is almost inevitable that adaptations will occur in the chronic absence of a particular gene. Furthermore, the consequences of developmental absence are almost certainly even more profound. The recent use of region-specific conditional knockouts, in which gene expression is altered only in particular anatomical regions and/or during particular times, has illustrated this point quite clearly and has simultaneously offered a solution. Perhaps the most relevant example to the field of anxiety comes from genetic studies of the 5-$HT_{1A}$ receptor.

The 5-$HT_{1A}$ receptor, as a target of many useful anxiolytic agonists, has been studied more than any other gene in the context of anxiety research. As might be expected, most knockout studies have reported increased anxiety-like behavior in the mutant animals. Seemingly, these data support a very simple, pharmacological view: 5-$HT_{1A}$ activation opposes anxiety. However, the recent work of Gross et al. (2002) has placed this interpretation in doubt. Using a knockout/rescue approach with regional and temporal specificity, Gross and colleagues demonstrated that the anxiety-related effect of the 5-$HT_{1A}$ receptor knockout was actually developmental. Specifically, expression limited to the hippocampus and cortex during early postnatal development was sufficient to counteract the anxious phenotype of the mutant, even though the receptor was still absent in adulthood (Gross et al. 2002). Thus, pharmacological and genetic/developmental effects of the 5-$HT_{1A}$ receptor on anxiety appear to be distinguishable. In conclusion, although there is no debating the usefulness of genetic manipulation, the interpretation of behavioral genetics studies must be thoughtful and cautious. There are undoubtedly many developmental insults, genetic and otherwise, that can alter the expression of anxiety-related behaviors. The dissection of a gene's role in anxiety must include the acute, adaptive/compensatory, and developmental effects. For further discussion of the genetic approaches to animal anxiety, the reader is referred to several recent reviews (Belzung 2001; Bilbo and Nelson 2001; Clement et al. 2002).

## *The Neurochemistry of Fear and Anxiety*

Much of our knowledge of the human neural substrates of fear and anxiety is derived from pioneering work with cat and rodent models (Davis 2002; LeDoux 2000). As techniques have advanced, our understanding of the anatomy, neurochemistry, and physiology of these responses has progressed. Particularly, the development of functional imaging techniques has allowed us to confirm that observations made in a number of animal species may also apply to humans.

The fear response necessarily begins with perception. Most sensory information is relayed via the thalamus to the sensory cortices, which are responsible for recognition and cognitive appraisal of threat. Two exceptions are olfactory input, which may reach the amygdala and entorhinal cortex directly, and visceral organ input, which proceeds from nuclei in the brain stem to the locus coeruleus (LC). The LC has several roles that are believed to be important in the regulation of anxiety. On one level, noradrenergic efferents appear to have a key role in regulating the peripheral sympathetic response; LC firing appears necessary for the generation of a physiological response to anxiogenic stimuli. Additionally, the LC, perhaps as a mediator of sensory cortex-to-thalamus information flow, also plays a role in directing attention toward salient, threatening stimuli. Notably, there is evidence that the release of CRH in the LC (see discussion later in this section) is necessary for the enhanced firing seen in high-anxiety conditions.

The amygdala is perhaps the best-studied, and most strongly implicated, brain structure in anxiety and fear (Davis 1992). Electrical stimulation of the amygdala produces fearlike behavioral and physiological responses in animals and increases the suggestive experience of fear in human subjects. Additionally, amygdala stimulation leads to corticosterone secretion and HPA axis activation, probably via outputs to the hypothalamus and the bed nucleus of the stria terminalis. It has been suggested that the amygdala may represent a sort of "master switch" of fear, which projects to a variety of areas (Davis 1992). This unitary-mediator hypothesis explains the constellation of behavioral and physiological responses that co-occur so consistently.

Insight into the pathophysiology of human anxiety comes from multiple studies suggesting that particular pharmacological agents often provoke symptoms of anxiety in susceptible individuals. These can be generally classified into those that reduce available oxygen, and appear to act on peripheral areas, and those that directly manipulate neurotransmitter systems. A number of pharmacological agents have been noted to increase anxiety (generally measured as panic symptoms in patients with panic disorder) in susceptible individuals (Sandford et al. 2000). In one sense these findings provide important clues in understanding the underlying neurobiology of anxiety disorders. However, the broad spectrum of agents that can result in anxiety symptoms makes it difficult to define one system, or pathway, in the brain most responsible for anxiety. It is possible, though, to broadly define the targets of compounds—which may have general mechanisms of action. In this regard, carbon dioxide, sodium lactate, and bicarbonate all act peripherally, resulting in increased respiration, heart rate, and other signs of sympathetic activity. This mechanism of action gave rise to a hypothesis by which patients with disorders of anxiety may interpret physiological changes (e.g., increased heart rate, increased respiration) as more serious and thus respond with increasing anxiety (Clark 1986). It may be that any peripherally adversive stimulus—especially one that stimulates sympathetic activity—has the potential to activate brain areas of prime importance in the formation of symptoms of anxiety. As a result of pharmacogical challenge studies, biochemical assays, neuroimaging, and studies of animal models, a number of centrally acting neurotransmitters—and their relevant neural circuits—are implicated in anxiety (see Table 5–6). These include norepinephrine, serotonin, GABA, neuropeptide Y, cholecystokinin, and substance P.

## The Noradrenergic System

Preclinical studies utilizing pharmacological manipulation and electrical stimulation have suggested the involvement of norepinephrine (NE) and the LC in anxiety- and fearlike behaviors. The strongest human evidence comes from studies of the anxiogenic and anxiolytic properties of centrally acting selective noradrenergic drugs, some of which exert these actions by increasing LC activity. The selective β-agonist isoproterenol induces anxiety and panic attacks in some individuals (Sandford et al. 2000); however, a direct attribution of the effects to enhanced central β-

**Table 5–6.** Primary neural systems implicated in anxiety

| Neural system | Evidence |
|---|---|
| Serotonin | Some serotonin agonists are anxiogenic (fenfluramine). |
| | SSRIs are effective for the treatment of anxiety. |
| | Tryptophan depletion can precipitate anxiety in some susceptible individuals. |
| Norepinephrine | The source of NE neurons in the brain (the locus coeruleus) is involved in control of arousal. |
| | Drugs that result in the increased release of NE often are anxiogenic (yohimbine). |
| CRH | CRH agonists increase anxiety in animal models. |
| | CRH antagonists decrease anxiety in animal models. |
| | CRH receptors are localized to many brain areas implicated in mediating anxiety symptoms. |
| | Cortisol secretion (downstream of CRH in the HPA) has been found to be abnormal in patients with anxiety. |
| GABA | Drugs that inhibit the function of GABA are anxiogenic (flumazenil). |
| | Drugs that potentiate the function of GABA are anxiolytic. |

*Note.* CRH = corticotropin-releasing hormone; GABA = γ-aminobutyric acid; HPA = hypothalamic-pituitary-adrenal axis; NE = norepinephrine; SSRI = selective serotonin reuptake inhibitor.

adrenergic receptor throughput is problematic, since this drug does not appear to cross the blood-brain barrier. More direct evidence comes from studies using the $\alpha_2$ antagonist yohimbine. $\alpha_2$ Adrenergic receptors are present on the cell bodies and terminals of NE neurons, where they regulate the firing rate and amount of NE released per nerve impulse, respectively. Thus, blockade of $\alpha_2$ receptors has been clearly established to increase the firing rate of

LC NE neurons, as well as NE release. In patients with panic disorder, compared with controls, yohimbine increases anxiety and the frequency of panic attacks—effects that have been attributed to enhanced LC firing and NE release (Charney et al. 1984; Uhde et al. 1984). Consistent with the behavioral data, yohimbine has also been shown to increase cardiovascular responses and plasma 3-methoxy-4-hydroxyphenyl glycol (MHPG, a metabolite of NE) and cortisol in panic patients relative to control subjects (Charney and Drevets 2002). Yohimbine has also been reported to produce a decrease in frontal blood flow in this patient population—effects that were not seen in control subjects (Woods et al. 1988). Additional pharmacological support for the role of the LC noradrenergic system in mediating anxiety symptoms comes from studies using clonidine, a centrally active $\alpha_2$ receptor agonist, which decreases LC firing and NE release. Thus, clonidine has been shown to have some efficacy in the treatment of anxiety and both spontaneous and induced panic disorder; interestingly, the beneficial effects subside over time, perhaps because of desensitization of the $\alpha_2$ receptor with continued stimulation by a direct agonist (Sandford et al. 2000). Furthermore, clonidine administration results in a greater degree of hypotension and larger reductions in plasma MHPG levels in patients with panic disorder relative to control subjects, raising the possibility of altered $\alpha_2$ receptor sensitivity in panic disorder (Charney and Drevets 2002). Consistent with such a contention, levels of NE and/or its metabolites measured in urine, blood, and CSF are elevated in patients with panic disorder, posttraumatic stress disorder (PTSD), specific phobias, social anxiety, and GAD, suggesting a dysregulation of the central and peripheral noradrenergic systems (reviewed in Charney and Drevets 2002).

It has been postulated that the inhibitory inputs from the frontal cortex to limbic regions may be disrupted in anxiety disorders, resulting in "unchecked" amygdala activity, resulting in an increase in anxiety. In this context, yohimbine decreases metabolism in the cortical brain areas in PTSD patients, whereas in control subjects cerebral metabolism is increased (Bremner et al. 1997a). Additionally, as mentioned, functional studies in panic disorder suggest dysregulated blood flow following administration of agents that result in panic symptoms. In addition to yohimbine (Woods et al. 1988), these studies have also been performed with lactate infusion (Stewart et al. 1988). The above findings in the cortex are consistent with the model by which cortical input to the amygdala provides a "top down" inhibition that is lacking in patients with anxiety disorders—and functions as a contributor to anxiety in patients. Also consistent is the general finding that while low levels of anxiety increase cortical metabolism in humans, high levels of anxiety have been shown to decrease it (Gur et al. 1987; Rodriguez et al. 1989). Thus, abnormal cerebral metabolism in anxiety may allow for abnormal amygdala activity. In this regard, recent preclinical work in rats has shown that stimulating specific areas of the medial PFC can increase the extinction of fear responses (Milad and Quirk 2002).

## The CRH and Stress Axes

The HPA axis is a major pathway by which stress exerts its effects on the brain and the rest of the body. This axis is also believed to have relevance in the development of anxiety and anxiety disorders (Bremner and Charney 2002). This pathway represents the primary "stress" pathway in humans, whereas the lateral ventricular nucleus of the hypothalamus releases corticotropin-releasing hormone/factor (CRH/F), which stimulates the production of adrenocorticotropic hormone (ACTH) by the pituitary. This hormone stimulates the production of cortisol by the adrenal gland. Cortisol is considered a primary stress hormone of the body, having varied effects on metabolism and neurovegetative behaviors of organisms, and effects on the functions of neurons and neuronal systems. Indeed, a primary receptor for cortisol—the glucocorticoid receptor—is localized to many brain regions important in the stress response, including a high density of receptors in the hippocampus.

CRH and its receptors are of major interest as a target for medications. In addition to being found in the pituitary, CRH receptors have been localized to the cortex, nuclei of the amygdala, LC, and regions of the hypothalamus. It has also been reported that CRH increases LC activity and that local injection of CRH into the LC increases behavioral responses consistent with increased anxiety (see Bremner and Charney 2002 for review). Furthermore, CRH antagonists have been repeatedly shown to have anxiolytic effects in animal models.

Supporting the notion that there is an overactivation of the HPA axis in anxiety are a number of clinical studies which suggest that patients with PTSD have a smaller hippocampal volume than matched control subjects (Bremner et al. 1995, 1997b; Charney and Bremner 1999; Gurvits et al. 1996; Stein et al. 1997)—a finding that is consistent with what has been observed in animal models of stress. To date, no quantitative neuroimaging studies have been performed in panic disorder, phobic disorder, or generalized anxiety disorder—but a single study does suggest the presence of abnormalities in the temporal lobe in patients with panic disorder (Ontiveros et al. 1989). Also, some studies suggest that there is an increase in cortisol release in response to stress in patients with PTSD and panic disorder. Furthermore, it is generally consistent that there is a chronic increase of CRH levels in patients with anxiety. Studies additionally suggest that there are changes in other mediators of the HPA axis, but these findings are generally less consistent.

A great deal of effort has gone into the development of antagonists to the type 1 and 2 CRH receptors. Thus, as mentioned, CRH agonism has proven anxiogenic in a variety of animal models, and CRH antagonists have demonstrated anxiolytic properties in rodent and primate studies (see Reul and Holsboer 2002), and even in a preliminary clinical study for depression (Zobel et al. 2000). Thus, drugs targeting the CRH family of ligands and receptors are perhaps the most promising new class of anxiolytics and antidepressants. The orally active CRH antagonist antalarmin significantly reduces fear and anxiety responses in nonhuman primates (Habib et al. 2000) and may be a candidate for further development. Several pharmaceutical companies have developed substituted pyrimidine small-molecule CRH antagonists (Arborelius et al. 1999). Trials are under way to examine the efficacy of these treatments in depression, and it is likely that complementary trials for the treatment of anxiety will also be undertaken. Targeting glucocorticoid receptors may also represent a possible mechanism to prevent stress mediated neuronal plasticity changes.

Thus, the evidence that stress, and the stress hormone corticosterone, may cause changes in hippocampal stucture, in hippocampal connections, and in gene and protein expression suggests that viable targets of anxiolytic agents are cellular signaling pathways involved in neuroplasticity and the maintenance of cellular resilience.

## The Serotonergic System

The raphe nuclei presumably play an important role in the serotonergic aspects of fear and anxiety. An excitatory projection from the LC to the dorsal raphe may be important in the serotonin release observed in the PFC, amygdala, and hypothalamus in response to anxiogenic stimuli. Additionally, projections from the dorsal raphe also extend to and inhibit the LC, suggesting a possible negative feedback mechanism. In this light, it is noteworthy that chronic, but not acute, SSRI administration suppresses LC firing in rats (Szabo and Blier 2002). The medial raphe may also have some opposing effects; there is some evidence that output from this nucleus to the dorsal hippocampus acts to increase stress resistance (Graeff et al. 1996), while LC projections to the medial raphe suppress its firing.

Evidence implicates the serotonergic system in the treatment of anxiety. Indeed, a number of medications useful for the treatment of anxiety have effects on serotonin neurotransmission. These include tricyclic antidepressant medications, selective serotonin-norepinephrine reuptake inhibitors, and monoamine oxidase inhibitors (MAOIs). However, as these medications take weeks to exert their full anxiolytic effects, it is unlikely that blocking the reuptake (and thus increasing synaptic levels) of either serotonin or norepinephrine selectively is responsible for their anxiolytic properties; rather, it is suspected that therapeutic effects are due to changes in gene expression and protein levels and ultimately changes in synaptic connections between neurons.

In spite of the large number of medications that target serotonin neurotransmission, consistent evidence implicating serotonin neurotransmission in the pathophysiology of anxiety is lacking. Indeed, as reviewed by Charney and Drevets (2002) and Charney and Bremner (1999), markers of the density and/or activity of serotonin reuptake transporters have been found to be increased, not changed, or decreased in panic disorders depending on the study and the experimental conditions. Furthermore, the application of pharmacological challenges to patients with anxiety disorders has been almost equally ambiguous. Thus, responses to serotonin precursors (L-tryptophan and 5-hydroxytryptophan) do not appear to be different between control subjects and those with panic disorder or to be anxiogenic. Challenges with other—more direct—serotonin agonists are also not always consistent. In this regard, the serotonin- releasing agent fenfluramine has been reported to be anxiogenic and to produce greater increases in plasma prolactin and cortisol in patients with panic disorder compared with control subjects. The serotonin agonist *m*-chloromethylpiperazine has been shown to increase anxiety and plasma cortisol in some studies of panic disorder patients compared with control subjects. Tryptophan depletion in anxiety disorders has also not been generally informative. Indirect evidence also implicates dysfuntion in PTSD, but more work needs to be done (Charney and Bremner 1999; Charney and Drevets 2002).

## The GABAergic System

The GABAergic system is the primary target for the acute treatment of anxiety. The benzodiazepines function by binding to a potentiator site on the $GABA_A$ receptor, increasing the amplitude and duration of inhibitory postsynaptic currents in response to GABA binding. Unfortunately, the benzodiazepines possess significant side effects, including sedation, cognitive impairment, and addictive liability. Our growing understanding of the GABAergic system may enable us to design more tolerable medications.

The $GABA_A$ receptor displays enormous heterogeneity, being composed of a combination of five subunits, of which there are at least 18 subtypes. The various receptors display variation in functional pharmacology, hinting at the multiple, finely tuned roles that inhibitory neurotransmission plays in brain function. The majority of $GABA_A$ receptors in the brain are targets of diazepam and other benzodiazepines. For this reason, there has been considerable interest in determining if the desirable and undesirable effects of benzodiazepines might be based on the differing receptor subtypes.

Much work has been conducted in this direction, particularly using gene knockout technology. For instance, mutation of the benzodiazepine-binding site of the $\alpha_1$ subunit in mice blocks the sedative, anticonvulsive, and amnesic, but not the anxiolytic, effects of diazepam. In contrast, the $\alpha_2$ subunit (expressed highly in the cortex and hippocampus) is necessary for diazepam anxiolysis and myorelaxation. Thus, there is now optimism that an $\alpha_2$ selective ligand will soon provide effective, acute treatment of anxiety disorders without the unfavorable side effect profile of current benzodiazepines. A compound with this preferential affinity has already been demonstrated to exert fewer sedative and depressant effects than diazepam in rat behavioral studies.

It is believed that novel medications under development will specifically target subtypes of GABA receptors, resulting in medications that are more specific for anxiety and that have fewer side effects. Additionally, current GABAergic drugs exert their actions almost immediately on initiation of treatment. Thus, there is optimism that novel medications of this class will also have immediate effects, thereby circumventing one of the primary problems of many current anxiolytic drugs—namely, the delay in onset of action. For more on the pharmacology of benzodiazepines, see the review by Mohler et al. (2002).

## Neuropeptide Y

Neuropeptide Y (NPY) and its receptors may also be important in the regulation of anxiety and stress. NPY is synthesized in the arcuate nucleus, which receives LC input. In a number of rodent models, NPY adminstration has anxiolytic and, at somewhat higher doses, sedative effects. Likewise, NPY antagonizes CRH-induced stress responses, and it suppresses LC firing when injected into the brain stem. There is some evidence to suggest that NPY is low under unstressed conditions but is stimulated as a counteradaptation to stress. Additionally, NPY projections to the central amygdala, nucleus accumbens, septum, periaqueductal gray (PAG), hippocampus, and other regions may also be involved in NPY anxiolysis. As our understanding of the differing roles of the various NPY receptors improves, NPY receptor agonists are likely to become a goal of anxiolytic development (see Kask et al. 2002 for further reading).

### Cholecystokinin

Cholecystokinin (CCK) is another neuropeptide with importance to anxiety disorder research. CCK-B receptor agonists reportedly have an anxiogenic effect in animals and are anxio- and panicogenic in normal subjects and panic disorder patients (patients are more sensitive). Likewise, suppression of CCK-B receptor expression or pharmacological antagonism blocks the acquisition of conditioned fear response. However, CCK antagonists have yet to succeed in clinical trials (Pande et al. 1999). Nevertheless, the CCK system remains an attractive target for drug development, particularly for panic disorders.

### Substance P

Substance P binds to the neurokinin-1 receptor (NK-1; also known as tachykinin-1 receptor). Although originally targeted as a nociception mediator, antagonists of this receptor may exert anxiolytic and antidepressant effects. There is also evidence for an anxiolytic effect of substance P when injected into the cholinergic nucleus basalis magnocellularis, suggesting region-specific effects. Furthermore, NK-1 activation in the hypothalamus inhibits CRH secretion. Several NK-1 antagonists appear anxiolytic in animal studies, and in a preliminary clinical trial, the antagonist MK 869 was found to be as effective as paroxetine in treating anxiety and depression. The interested reader is referred to the review by Stout et al. (2001) for additional discussion about the potential usefulness of NK-1 antagonists.

## *Intracellular Targets for Anxiety Disorders*

While genetic and pharmacological studies have focused predominantly on the receptors, transporters, and metabolic enzymes of neurotransmitters already implicated in anxiety, there also exists a more diverse group of gene knockouts with anxiety-related phenotypes. A number of studies have investigated anxiety-like behavior in knockouts of key intracellular signaling enzymes, including CaMKIIα, adenylate cyclase type VIII, and protein kinase C ε and γ. These studies raise the possibility that pharmacological manipulation of signaling cascades *within* cells might be exploited in the development of new anxiolytics.

PKC provides a particularly good example. The knockout of either the ε or γ isoform confers resistance to anxiety and potentiates the action of benzodiazepines. Furthermore, PKC inhibition may have anxiolytic effects in humans: in a study of psychosocial functioning in women on long-term tamoxifen, which possesses both PKC- and estrogen-inhibiting properties, there was a trend toward reduced anxiety in the treatment group (Fallowfield et al. 2001). Likewise, valproic acid, an anticonvulsant and mood stabilizer that reduces expression of several PKC isoforms in preclinical models, may have efficacy in the treatment of panic disorder (Baetz and Bowen 1998) and PTSD (Clark et al. 1999). The specific roles of PKC isozymes in anxiety behavior require study, but specific inhibitors may someday serve as anxiolytics or anxiolytic potentiators.

Inositol, the building block of the phosphoinositide intracellular signaling pathway, has been examined as a potential anxiolytic. In clinical trials, inositol has reported efficacy in both panic disorder (Benjamin et al. 1995; Palatnik et al. 2001) and depression (Levine et al. 1995); animal data are also favorable (Einat and Belmaker 2001). Because of inositol's status as a dietary supplement, there is little financial backing for studies of its efficacy and safety. The mechanism of action of inositol is not entirely clear, but it is believed to facilitate phosphoinositide signaling–coupled neurotransmission.

## Conclusion

In this chapter, we have attempted to highlight recent neurobiological findings, which are generating considerable excitement about the development of novel agents for the treatment of mood and anxiety disorders (see Appendix 2 at end of this chapter). Most notably, there is a considerable body of evidence both conceptually and experimentally suggesting that impairments in neuroplasticity and cellular resilience are demonstrable in certain mood and anxiety disorders; thus, in addition to a variety of neurochemical changes, many patients also have pronounced *structural alterations* (e.g., reduced spine density, neurite retraction, overall neuropil reductions, and/or volumetric changes on neuroimmaging measures) in critical neuronal circuits. For these disorders, optimal treatment may be attained only by providing both trophic and neurochemical support; the trophic support would be envisioned as enhancing and maintaining normal synaptic connectivity, thereby allowing the chemical signal to reinstate the optimal functioning of critical circuits necessary for normal affective functioning. It is noteworthy that many novel strategies that are currently being investigated for these disorders—including CRH antagonists, GR antagonists, antiglutamatergic agents, and phosphodiesterase inhibitors—can be conceptualized as "plasticity enhancers," which would be expected to exert trophic effects in addition to their effects on specifical neurochemical systems (Figure 5–4; Appendix 2).

It is hoped that our ever-expanding knowledge of the neurobiology of anxiety and anxiety disorders will yield entirely new pharmacological approaches to the treatment of these disorders. Despite some concerns with their validity, animal models continue to provide a useful screening mechanism for the development of novel anxiolytics. There are at present a number of promising new targets, and a new generation of drugs directed at some of these may come to the market in the coming years (Appendix 2). Traditionally, the neurochemical systems targeted in the treatment of anxiety disorders have been GABA, serotonin, and norepinephrine. Continuing development of drugs modulating these transmitters will probably yield modestly more effective and/or better tolerated medications. However, the results achieved with more innovative targets promise both markedly better drugs and a more sophisticated understanding of pathological anxiety.

The evidence also suggests that somewhat akin to the treatment of conditions like hypertension and diabetes, early and potentially sustained treatment may be necessary to adequately prevent many of the deleterious long-term sequelae associated with certain mood and anxiety disorders. While data suggest that hippocampal atrophy in depression may be related to illness duration (Sheline et al. 1999), it is at present not clear if the volumetric and cellular changes that have been observed in other brain areas (most notably frontal cortex) are related to affective episodes per se. Indeed, some studies have observed reduced gray matter volumes and enlarged ventricles in mood disorder patients at first or early onset (Hirayasu et al. 1999; Strakowski et al. 1993). At this point, it is unclear if the regional cellular atrophy in mood disorders occurs because of the magnitude and duration of biochemical perturbations, an enhanced vulnerability to the deleterious effects of these perturbations (due to genetic factors and/or early life events), or a combination thereof. Likewise, the hippocampal volumetric changes observed in subjects with PTSD—and the altered cortical activation in subjects with anxiety disorders—may be due to a genetic diathesis or hormonal (HPA/CRH) alterations or may be epiphenomena of other occurrences. In this context, a growing body of data has demonstrated that neonatal stress can have a major impact on brain development, in particular by bringing about persistent changes in CRH-containing neurons, the HPA axis, the serotonergic system, the noradrenergic system, and the sympathetic nervous system (Graham et al. 1999). The possibility that these neurochemical alterations produce a state of neuroendangerment, as described in this chapter, that contributes to the subsequent development of morphological brain changes in adulthood requires further investigation.

Furthermore, while there are undoubtedly genetic contributions (both those of susceptibility and those of protection) to the impact of neonatal stresses on brain development, it is noteworthy that recent studies have also demonstrated nongenomic transmission across generations of not only maternal behavior but also stress responses (Francis et al. 1999). This has clear parallels in clinical populations. For example, environmental events (e.g., early childhood stressors) correlate with the development of psychiatric disorders in adults (Heim and Nemeroff 2001). Indeed—as witnessed by multiple studies of discordant monozygotic twins when one has the disorder and the other does not—epigenetic mechanisms must be operative to control behavior in genetically identical populations (Gottesman et al. 1982). A critical question involves the mechanism by which early life events regulate long-term changes in behavior and sustained differences in gene expression.

In most areas of the brain, neurons are not replaced. Thus, permanent and semipermanent modifications that occur in early life, which alter gene transcription, could have downstream effects that may be temporally distant from the initial event. These nongenetic mechanisms of gene regulation are termed *epigenetic* and likely play a role in the formation of cellular memory and the modulation of neural circuitry in a manner that alters lifetime cellular and behavioral responses in an organism. The true extent of the dynamic mechanisms responsible is unknown and is an active area of research. However, it is known to involve the interplay of transcription factors interacting with covalent DNA modifications, such as cytosine methylation, and the accessibility of DNA that is regulated by histone acetylation (Geiman and Robertson 2002; Petronis 2001). It is likely that these mechanisms are involved in modulating how previous experience may regulate subsequent behavioral responses.

In conclusion, emerging results from a variety of clinical and preclinical experimental and naturalistic paradigms suggest that a reconceptualization about the pathophysiology, course, and optimal long-term treatment of severe mood and anxiety disorders may be warranted. An increasing number of strategies are being investigated to develop small-molecule agents to enhance neuroplasticity and cellular resilience. This progress holds much promise for the development of novel therapeutics for the long-term treatment of severe, refractory mood disorders and for improving the lives of millions.

## References

Altamura CA, Mauri MC, Ferrara A, et al: Plasma and platelet excitatory amino acids in psychiatric disorders. Am J Psychiatry 150(11): 1731–1733, 1993

Altshuler LL, Casanova MF, Goldberg TE, et al: The hippocampus and parahippocampus in schizophrenia, suicide, and control brains. Arch Gen Psychiatry 47:1029–1034, 1990

Arborelius L, Owens MJ, Plotsky PM, et al: The role of corticotropin-releasing factor in depression and anxiety disorders. J Endocrinol 160: 1–12, 1999

Asanuma M, Nishibayashi S, Iwata E, et al: Alterations of cAMP response element–binding activity in the aged rat brain in response to administration of rolipram, a cAMP-specific phosphodiesterase inhibitor. Brain Res Mol Brain Res 41(1–2):210–215, 1996

Auer DP, Putz B, Kraft E, et al: Reduced glutamate in the anterior cingulate cortex in depression: an in vivo proton magnetic resonance spectroscopy study. Biol Psychiatry 47(4):305–313, 2000

Baetz M, Bowen RC: Efficacy of divalproex sodium in patients with panic disorder and mood instability who have not responded to conventional therapy. Can J Psychiatry 43(1):73–77, 1998

Bale TL, Picetti R, Contarino A, et al: Mice deficient for both corticotropin-releasing factor receptor 1 (CRFR1) and CRFR2 have an impaired stress response and display sexually dichotomous anxiety-like behavior. J Neurosci 22:193–199, 2002

Baltuch GH, Couldwell WT, Villemure JG, et al: Protein kinase C inhibitors suppress cell growth in established and low-passage glioma cell lines: a comparison between staurosporine and tamoxifen. Neurosurgery 33(3):495–501; discussion 501, 1993

Barros M, Tomaz C: Non-human primate models for investigating fear and anxiety. Neurosci Biobehav Rev 26(2):187–201, 2002

Baumann B, Danos P, Krell D, et al: Unipolar-bipolar dichotomy of mood disorders is supported by noradrenergic brainstem system morphology. J Affect Disord 54(1–2):217–224, 1999

Bebchuk JM, Arfken CL, Dolan-Manji S, et al: A preliminary investigation of a protein kinase C inhibitor in the treatment of acute mania. Arch Gen Psychiatry 57(1):95–97, 2000

Belanoff JK, Flores BH, Kalezhan M, et al: Rapid reversal of psychotic depression using mifepristone. J Clin Psychopharmacol 21(5):516–521, 2001

Belzung C: The genetic basis of the pharmacological effects of anxiolytics: a review based on rodent models. Behav Pharmacol 12(6–7):451–460, 2001

Benes FM, Kwok EW, Vincent SL, et al: A reduction of nonpyramidal cells in sector CA2 of schizophrenics and manic depressives. Biol Psychiatry 44:88–97, 1998

Benes FM, Todtenkopf MS, Logiotatos P, et al: Glutamate decarboxylase(65)–immunoreactive terminals in cingulate and prefrontal cortices of schizophrenic and bipolar brain. J Chem Neuroanat 20:259–269, 2000

Benes FM, Vincent SL, Todtenkopf M: The density of pyramidal and nonpyramidal neurons in anterior cingulate cortex of schizophrenic and bipolar subjects. Biol Psychiatry 50(6):395–406, 2001

Benjamin J, Levine J, Fux M, et al: Double-blind, placebo-controlled, crossover trial of inositol treatment for panic disorder. Am J Psychiatry 152(7):1084–1086, 1995

Berk M: Lamotrigine and the treatment of mania in bipolar disorder. Eur Neuropsychopharmacol 9 (suppl 4):S119–S123, 1999

Berman RM, Cappiello A, Anand A, et al: Antidepressant effects of ketamine in depressed patients. Biol Psychiatry 47(4):351–354, 2000

Bertolino A, Crippa D, di Dio S, et al: Rolipram versus imipramine in inpatients with major, "minor" or atypical depressive disorder: a double-blind double-dummy study aimed at testing a novel therapeutic approach. Int Clin Psychopharmacol 3(3):245–253, 1988

Beyer JL, Krishnan KR: Volumetric brain imaging findings in mood disorders. Bipolar Disord 4(2):89–104, 2002

Bezchlibnyk Y, Young LT: The neurobiology of bipolar disorder: focus on signal transduction pathways and the regulation of gene expression. Can J Psychiatry 47(2):135–148, 2002

Bhalla US, Iyengar R: Emergent properties of networks of biological signaling pathways. Science 283(5400):381–387, 1999

Bilbo SD, Nelson RJ: Behavioral phenotyping of transgenic and knockout animals: a cautionary tale. Lab Anim (NY) 30(1):24–29, 2001

Blednov YA, Stoffel M, Chang SR, et al: GIRK2 deficient mice: evidence for hyperactivity and reduced anxiety. Physiol Behav 74:109–117, 2001a

Blednov YA, Stoffel M, Chang SR, et al: Potassium channels as targets for ethanol: studies of G-protein-coupled inwardly rectifying potassium channel 2 (GIRK2) null mutant mice. J Pharmacol Exp Ther 298: 521–530, 2001b

Bobon D, Breulet M, Gerard-Vandenhove MA, et al: Is phosphodiesterase inhibition a new mechanism of antidepressant action? A double blind double-dummy study between rolipram and desipramine in hospitalized major and/or endogenous depressives. Eur Arch Psychiatry Neurol Sci 238(1):2–6, 1988

Bonni A, Brunet A, West AE, et al: Cell survival promoted by the Ras-MAPK signaling pathway by transcription-dependent and -independent mechanisms. Science 286(5443):1358–1362, 1999

Bouras C, Kovari E, Hof PR, et al: Anterior cingulate cortex pathology in schizophrenia and bipolar disorder. Acta Neuropathol (Berl) 102(4): 373–379, 2001

Bourne HR, Nicoll R: Molecular machines integrate coincident synaptic signals. Cell 72(suppl):65–75, 1993

Bouwknecht JA, Hijzen TH, van der Gugten J, et al: Absence of 5-$HT_{1B}$ receptors is associated with impaired impulse control in male 5-$HT_{1B}$ knockout mice. Biol Psychiatry 49:557–568, 2001

Bowers BJ, Collins AC, Tritto T, et al: Mice lacking PKC gamma exhibit decreased anxiety. Behav Genet 30:111–121, 2000

Bowers BJ, Elliott KJ, Wehner JM: Differential sensitivity to the anxiolytic effects of ethanol and flunitrazepam in PKC gamma null mutant mice. Pharmacol Biochem Behav 69:99–110, 2001

Bowley MP, Drevets WC, Ongur D, et al: Low glial numbers in the amygdala in major depressive disorder. Biol Psychiatry 52(5):404–412, 2002

Boyer PA, Skolnick P, Fossom LH: Chronic administration of imipramine and citalopram alters the expression of NMDA receptor subunit mRNAs in mouse brain: a quantitative in situ hybridization study. J Mol Neurosci 10(3):219–233, 1998

Bremner JD, Charney DS: Neural circuits in fear and anxiety, in Textbook of Anxiety Disorders. Edited by Stein DJ, Hollander E. Washington, DC, Amercian Psychiatric Publishing, 2002, pp 43–56

Bremner JD, Innis RB, Ng CK, et al: Positron emission tomography measurement of cerebral metabolic correlates of yohimbine administration in combat-related posttraumatic stress disorder. Arch Gen Psychiatry 54(3):246–254, 1997a

Bremner JD, Randall P, Vermetten E, et al: Magnetic resonance imaging–based measurement of hippocampal volume in posttraumatic stress disorder related to childhood physical and sexual abuse—a preliminary report. Biol Psychiatry 41(1):23–32, 1997b

Bremner JD, Randall P, Scott TM, et al: MRI-based measurement of hippocampal volume in patients with combat-related posttraumatic stress disorder. Am J Psychiatry 152(7):973–981, 1995

Brunner D, Buhot MC, Hen R, et al: Anxiety, motor activation, and maternal-infant interactions in 5$HT_{1B}$ knockout mice. Behav Neurosci 113:587–601, 1999

Calabrese JR, Bowden CL, Sachs GS, et al: A double-blind placebo-controlled study of lamotrigine monotherapy in outpatients with bipolar I depression. Lamictal 602 Study Group. J Clin Psychiatry 60(2):79–88, 1999

Cameron HA, McKay RD: Restoring production of hippocampal neurons in old age. Nat Neurosci 2(10):894–897, 1999

Cases O, Seif I, Grimsby J, et al: Aggressive behavior and altered amounts of brain serotonin and norepinephrine in mice lacking MAOA. Science 268:1763–1766, 1995

Castillo M, Kwock L, Courvoisie H, et al: Proton MR spectroscopy in children with bipolar affective disorder: preliminary observations. AJNR Am J Neuroradiol 21(5):832–838, 2000

Charney DS, Bremner JD: The neurobiology of anxiety disorders, in Neurobiology of Mental Illness. Edited by Charney DS, Nestler EJ, Bunney BS. New York, Oxford University Press, 1999, pp 494–517

Charney DS, Drevets WC: The neurobiological basis of anxiety disorders, in Neuropsychopharmacology: The Fifth Generation of Progress. Edited by Davis KL, Charney DS, Coyle JT, et al. Philadelphia, PA, Lippincott Williams & Wilkins, 2002, pp 901–930

Charney DS, Heninger GR, Breier A: Noradrenergic function in panic anxiety: effects of yohimbine in healthy subjects and patients with agoraphobia and panic disorder. Arch Gen Psychiatry 41(8):751–763, 1984

Chemerinski E, Robinson RG: The neuropsychiatry of stroke. Psychosomatics 41(1):5–14, 2000

Chen B, Dowlatshahi D, MacQueen GM, et al: Increased hippocampal BDNF immunoreactivity in subjects treated with antidepressant medication. Biol Psychiatry 50(4):260–265, 2001

Chen C, Rainnie DG, Greene RW, et al: Abnormal fear response and aggressive behavior in mutant mice deficient for alpha-calcium-calmodulin kinase II. Science 266:291–294, 1994

Chen DF, Tonegawa S: Why do mature CNS neurons of mammals fail to re-establish connections following injury—functions of bcl-2. Cell Death Differ 5(10):816–822, 1998

Chen DF, Schneider GE, Martinou JC, et al: Bcl-2 promotes regeneration of severed axons in mammalian CNS. Nature 385(6615):434–439, 1997

Chen G, Manji HK, Hawver DB, et al: Chronic sodium valproate selectively decreases protein kinase C alpha and epsilon in vitro. J Neurochem 63(6):2361–2364, 1994

Chen G, Huang LD, Jiang YM, et al: The mood-stabilizing agent valproate inhibits the activity of glycogen synthase kinase–3. J Neurochem 72(3):1327–1330, 1999a

Chen G, Zeng WZ, Yuan PX, et al: The mood-stabilizing agents lithium and valproate robustly increase the levels of the neuroprotective protein bcl-2 in the CNS. J Neurochem 72(2):879–882, 1999b

Chen G, Rajkowska G, Du F, et al: Enhancement of hippocampal neurogenesis by lithium. J Neurochem 75(4):1729–1734, 2000

Chen G, Huang LD, Zeng WZ, et al: Mood stabilizers regulate cytoprotective and mRNA-binding proteins in the brain: long-term effects on cell survival and transcript stability. Int J Neuropsychopharmacol 4(1):47–64, 2001

Chen RH, Ding WV, McCormick F: Wnt signaling to beta-catenin involves two interactive components: glycogen synthase kinase–3beta inhibition and activation of protein kinase C. J Biol Chem 275(23):17894–17899, 2000

Chen SJ, Sweatt JD, Klann E: Enhanced phosphorylation of the postsynaptic protein kinase C substrate RC3/neurogranin during long-term potentiation. Brain Res 749(2):181–187, 1997

Chierzi S, Strettoi E, Cenni MC, et al: Optic nerve crush: axonal responses in wild-type and bcl-2 transgenic mice. J Neurosci 19(19):8367–8376, 1999

Clark DM: A cognitive approach to panic. Behav Res Ther 24(4):461–470, 1986

Clark RD, Canive JM, Calais LA, et al: Divalproex in posttraumatic stress disorder: an open-label clinical trial. J Trauma Stress 12(2):395–401, 1999

Clement Y, Calatayud F, Belzung C: Genetic basis of anxiety-like behaviour: a critical review. Brain Res Bull 57(1):57–71, 2002

Collinson N, Kuenzi FM, Jarolimek W, et al: Enhanced learning and memory and altered GABAergic synaptic transmission in mice lacking the alpha 5 subunit of the $GABA_A$ receptor. J Neurosci 22:5572–55780, 2002

Conn PJ, Sweatt JD: Protein kinase C in the nervous system, in Protein Kinase C. Edited by Kuo JF. New York, Oxford University Press, 1994, pp 199–235

Contarino A, Dellu F, Koob GF, et al: Reduced anxiety-like and cognitive performance in mice lacking the corticotropin-releasing factor receptor 1. Brain Res 835:1–9, 1999

Contarino A, Dellu F, Koob GF, et al: Dissociation of locomotor activation and suppression of food intake induced by CRF in CRFR1-deficient mice. Endocrinology 141:2698–2702, 2000

Cotter D, Mackay D, Landau S, et al: Reduced glial cell density and neuronal size in the anterior cingulate cortex in major depressive disorder. Arch Gen Psychiatry 58(6):545–553, 2001

Coull MA, Lowther S, Katona CL, et al: Altered brain protein kinase C in depression: a post-mortem study. Eur Neuropsychopharmacol 10(4):283–288, 2000

Coyle JT, Schwarcz R: Mind glue: implications of glial cell biology for psychiatry. Arch Gen Psychiatry 57(1):90–93, 2000
Cross DA, Culbert AA, Chalmers KA, et al: Selective small-molecule inhibitors of glycogen synthase kinase–3 activity protect primary neurones from death. J Neurochem 77(1):94–102, 2001
Czeh B, Michaelis T, Watanabe T, et al: Stress-induced changes in cerebral metabolites, hippocampal volume, and cell proliferation are prevented by antidepressant treatment with tianeptine. Proc Natl Acad Sci U S A 98(22):12796–12801, 2001
Davis M: The role of the amygdala in fear and anxiety. Annu Rev Neurosci 15:353–375, 1992
Davis M: Neural circuitry of anxiety and stress disorders, in Neuropsychopharmacology: The Fifth Generation of Progress. Edited by Davis KL, Charney DS, Coyle JT, et al. Philadelphia, PA, Lippincott Williams & Wilkins, 2002, pp 931–951
Dean B, Pavey G, McLeod M, et al: A change in the density [$^3$H]flumazenil, but not [$^3$H]muscimol, binding in Brodmann's area 9 from subjects with bipolar disorder. J Affect Disord 66(2–3):147–158, 2001
DeVries AC, Joh HD, Bernard O, et al: Social stress exacerbates stroke outcome by suppressing Bcl-2 expression. Proc Natl Acad Sci U S A 98(20):11824–11828, 2001
Dixon JF, Hokin LE: Lithium acutely inhibits and chronically up-regulates and stabilizes glutamate uptake by presynaptic nerve endings in mouse cerebral cortex. Proc Natl Acad Sci U S A 95(14):8363–8368, 1998
Drevets WC: Neuroimaging studies of mood disorders. Biol Psychiatry 48(8):813–829, 2000
Drevets WC: Neuroimaging and neuropathological studies of depression: implications for the cognitive-emotional features of mood disorders. Curr Opin Neurobiol 11(2):240-9, 2001
Drevets WC, Price JL, Simpson JR Jr, et al: Subgenual prefrontal cortex abnormalities in mood disorders. Nature 386(6627):824–827, 1997
D'Sa C, Duman R: Antidepressants and neuroplasticity. Bipolar Disord 4(3):183–194, 2002
Dulawa SC, Grandy DK, Low MJ, et al: Dopamine D4 receptor knockout mice exhibit reduced exploration of novel stimuli. J Neurosci 19:9550–9556, 1999
Duman R: Synaptic plasticity and mood disorders. Mol Psychiatry 7 (suppl 1):S29–S34, 2002
Duman RS, Malberg J, Thome J: Neural plasticity to stress and antidepressant treatment. Biol Psychiatry 46(9):1181–1191, 1999

Dyke HJ, Montana JG: Update on the therapeutic potential of PDE4 inhibitors. Expert Opin Investig Drugs 11(1):1–13, 2002

Einat H, Belmaker RH: The effects of inositol treatment in animal models of psychiatric disorders. J Affect Disord 62(1–2):113–121, 2001

Eriksson PS, Perfilieva E, Bjork-Eriksson T, et al: Neurogenesis in the adult human hippocampus. Nat Med 4(11):1313–1317, 1998

Fallowfield L, Fleissig A, Edwards R, et al: Tamoxifen for the prevention of breast cancer: psychosocial impact on women participating in two randomized controlled trials. J Clin Oncol 19(7):1885–1892, 2001

Falzone TL, Gelman DM, Young JI, et al: Absence of dopamine D4 receptors results in enhanced reactivity to unconditioned, but not conditioned, fear. Eur J Neurosci 15:158–164, 2002

Filliol D, Ghozland S, Chluba J, et al: Mice deficient for delta- and mu-opioid receptors exhibit opposing alterations of emotional responses. Nat Genet 25:195–200, 2000

Finkbeiner S: CREB couples neurotrophin signals to survival messages. Neuron 25(1):11–14, 2000

Fleischhacker WW, Hinterhuber H, Bauer H, et al: A multicenter double-blind study of three different doses of the new cAMP-phosphodiesterase inhibitor rolipram in patients with major depressive disorder. Neuropsychobiology 26(1–2):59–64, 1992

Francis D, Diorio J, Liu D, et al: Nongenomic transmission across generations of maternal behavior and stress responses in the rat. Science 286(5442):1155–1158, 1999

Friedman E, Hoau Yan W, et al: Altered platelet protein kinase C activity in bipolar affective disorder, manic episode. Biol Psychiatry 33(7): 520–525, 1993

Fujimaki K, Morinobu S, Duman RS: Administration of a cAMP phosphodiesterase 4 inhibitor enhances antidepressant-induction of BDNF mRNA in rat hippocampus. Neuropsychopharmacology 22(1):42–51, 2000

Gass P, Kretz O, Wolfer DP, et al: Genetic disruption of mineralocorticoid receptor leads to impaired neurogenesis and granule cell degeneration in the hippocampus of adult mice. EMBO Rep 1(5):447–451, 2000

Geiman TM, Robertson KD: Chromatin remodeling, histone modifications, and DNA methylation—how does it all fit together? J Cell Biochem 87(2):117–125, 2002

Goggi J, Pullar IA, Carney SL, et al: Modulation of neurotransmitter release induced by brain-derived neurotrophic factor in rat brain striatal slices in vitro. Brain Res 941(1–2):34–42, 2002

Gogos JA, Morgan M, Luine V, et al: Catechol-O-methyltransferase-deficient mice exhibit sexually dimorphic changes in catecholamine levels and behavior. Proc Natl Acad Sci U S A 95:9991–9996, 1998

Gold PW, Chrousos GP: Organization of the stress system and its dysregulation in melancholic and atypical depression: high vs low CRH/NE states. Mol Psychiatry 7(3):254–275, 2002

Gold PW, Drevets WC, Charney DS: New insights into the role of cortisol and the glucocorticoid receptor in severe depression. Biol Psychiatry 52(5):381–385, 2002

Gottesman II, Shields J, Hanson DR: Schizophrenia, the Epigenetic Puzzle. Cambridge, UK, Cambridge University Press, 1982

Gould E, Tanapat P, Rydel T, et al: Regulation of hippocampal neurogenesis in adulthood. Biol Psychiatry 48(8):715–720, 2000

Gould TD, Manji HK: Signaling networks in the pathophysiology and treatment of mood disorders. J Psychosom Res 53(2):687–697, 2002a

Gould TD, Manji HK: The wnt signaling pathway in bipolar disorder. Neuroscientist 8(5):497–511, 2002b

Gould TD, Chen G, Manji HK: Mood stabilizer psychopharmacology. Clin Neurosci Res 2(2):193–212, 2002

Graeff FG, Guimaraes FS, De Andrade TG, et al: Role of 5-HT in stress, anxiety, and depression. Pharmacol Biochem Behav 54(1):129–141, 1996

Graham YP, Heim C, Goodman SH, et al: The effects of neonatal stress on brain development: implications for psychopathology. Dev Psychopathol 11(3):545–565, 1999

Griebel G, Misslin R, Vogel E, et al: Behavioral effects of rolipram and structurally related compounds in mice: behavioral sedation of cAMP phosphodiesterase inhibitors. Pharmacol Biochem Behav 39(2):321–323, 1991

Grimes CA, Jope RS: CREB DNA binding activity is inhibited by glycogen synthase kinase–3 beta and facilitated by lithium. J Neurochem 78(6):1219–1232, 2001

Grootendorst J, de Kloet ER, Dalm S, et al: Reversal of cognitive deficit of apolipoprotein E knockout mice after repeated exposure to a common environmental experience. Neuroscience 108:237–247, 2001

Gross C, Zhuang X, Stark K, et al: Serotonin$_{1A}$ receptor acts during development to establish normal anxiety-like behaviour in the adult. Nature 416(6879):396–400, 2002

Gur RC, Gur RE, Resnick SM, et al: The effect of anxiety on cortical cerebral blood flow and metabolism. J Cereb Blood Flow Metab 7(2):173–177, 1987

Gurvits TV, Shenton ME, Hokama H, et al: Magnetic resonance imaging study of hippocampal volume in chronic, combat-related posttraumatic stress disorder. Biol Psychiatry 40(11):1091–1099, 1996

Habib KE, Weld KP, Rice KC, et al: Oral administration of a corticotropin-releasing hormone receptor antagonist significantly attenuates behavioral, neuroendocrine, and autonomic responses to stress in primates. Proc Natl Acad Sci U S A 97(11):6079–6084, 2000

Hahn CG, Friedman E: Abnormalities in protein kinase C signaling and the pathophysiology of bipolar disorder. Bipolar Disord 1(2):81–86, 1999

Hall AC, Brennan A, Goold RG, et al: Valproate regulates GSK-3-mediated axonal remodeling and synapsin I clustering in developing neurons. Mol Cell Neurosci 20(2):257–270, 2002

Haydon PG: GLIA: listening and talking to the synapse. Nat Rev Neurosci 2(3):185–193, 2001

He M, Sibille E, Benjamin D, et al: Differential effects of 5-$HT_{1A}$ receptor deletion upon basal and fluoxetine-evoked 5-HT concentrations as revealed by in vivo microdialysis. Brain Res 902:11–17, 2001

Hebenstreit GF, Fellerer K, Fichte K, et al: Rolipram in major depressive disorder: results of a double-blind comparative study with imipramine. Pharmacopsychiatry 22(4):156–160, 1989

Heim C, Nemeroff CB: The role of childhood trauma in the neurobiology of mood and anxiety disorders: preclinical and clinical studies. Biol Psychiatry 49(12):1023–1039, 2001

Heisler LK, Chu HM, Brennan TJ, et al: Elevated anxiety and antidepressant-like responses in serotonin 5-$HT_{1A}$ receptor mutant mice. Proc Natl Acad Sci U S A 95:15049–15054, 1998

Hirayasu Y, Shenton ME, Salisbury DF, et al: Subgenual cingulate cortex volume in first-episode psychosis. Am J Psychiatry 156(7):1091–1093, 1999

Hodge CW, Mehmert KK, Kelley SP, et al: Supersensitivity to allosteric $GABA_A$ receptor modulators and alcohol in mice lacking PKC epsilon. Nat Neurosci 2:997–1002, 1999

Holm KH, Cicchetti F, Bjorklund L, et al: Enhanced axonal growth from fetal human bcl-2 transgenic mouse dopamine neurons transplanted to the adult rat striatum. Neuroscience 104(2):397–405, 2001

Holsboer F: The corticosteroid receptor hypothesis of depression. Neuropsychopharmacology 23(5):477–501, 2000

Horowski R, Sastre-Y-Hernandez M: Clinical effects of the neurotropic selective cAMP phosphodiesterase inhibitor rolipram in depressed patients: global evaluation of the preliminary reports. Current Therapeutic Research 38(1):23–29, 1985

Huang Z, Ducharme Y, Macdonald D, et al: The next generation of PDE4 inhibitors. Curr Opin Chem Biol 5(4):432–438, 2001

Hurley SC: Lamotrigine update and its use in mood disorders. Ann Pharmacother 36(5):860–873, 2002

Ikegaya Y, Yamada M, Fukuda T, et al: Aberrant synaptic transmission in the hippocampal CA3 region and cognitive deterioration in protein-repair enzyme-deficient mice. Hippocampus 11:287–298, 2001

Jacobs BL, Praag H, Gage FH: Adult brain neurogenesis and psychiatry: a novel theory of depression. Mol Psychiatry 5(3):262–269, 2000

Johannessen CU, Petersen D, Fonnum F, et al: The acute effect of valproate on cerebral energy metabolism in mice. Epilepsy Res 47(3): 247–256, 2001

Johansson B, Halldner L, Dunwiddie TV, et al: Hyperalgesia, anxiety, and decreased hypoxic neuroprotection in mice lacking the adenosine A1 receptor. Proc Natl Acad Sci U S A 98:9407–9412, 2001

Kash SF, Tecott LH, Hodge C, et al: Increased anxiety and altered responses to anxiolytics in mice deficient in the 65-kDa isoform of glutamic acid decarboxylase. Proc Natl Acad Sci U S A 96:1698–1703, 1999

Kask A, Harro J, von Horsten S, et al: The neurocircuitry and receptor subtypes mediating anxiolytic-like effects of neuropeptide Y. Neurosci Biobehav Rev 26(3):259–283, 2002

Kempermann G: Regulation of adult hippocampal neurogenesis—implications for novel theories of major depression. Bipolar Disord 4(1): 17–33, 2002

Kishimoto T, Radulovic J, Radulovic M, et al: Deletion of crhr2 reveals an anxiolytic role for corticotropin-releasding hormone receptor-2. Nat Genet 24:415–419, 2000

Klein PS, Melton DA: A molecular mechanism for the effect of lithium on development. Proc Natl Acad Sci U S A 93:8455–8459, 1996

Klemfuss H: Rhythms and the pharmacology of lithium. Pharmacol Ther 56(1):53–78, 1992

Korte SM, De Kloet ER, Buwalda B, et al: Antisense to the glucocorticoid receptor in hippocampal dentate gyrus reduces immobility in forced swim test. Eur J Pharmacol 301(1–3):19–25, 1996

Koster A, Montkowski A, Schulz S, et al: Targeted disruption of the orphanin FQ/nociceptin gene increases stress susceptibility and impairs stress adaptation in mice. Proc Natl Acad Sci U S A 96:10444–10449, 1999

Krezel W, Dupont S, Krust A, et al: Increased anxiety and synaptic plasticity in estrogen receptor beta-deficient mice. Proc Natl Acad Sci U S A 98:12278–12282, 2001

Krystal JH, Sanacora G, Blumberg H, et al: Glutamate and GABA systems as targets for novel antidepressant and mood-stabilizing treatments. Mol Psychiatry 7 (suppl 1):S71–S80, 2002

Kumar A, Miller D, Ewbank D, et al: Quantitative anatomic measures and comorbid medical illness in late-life major depression. Am J Geriatr Psychiatry 5(1):15–25, 1997

Kustova Y, Sei Y, Morse HC Jr, et al: The influence of a targeted deletion of the IFN gamma gene on emotional behaviors. Brain Behav Immun 12:308–324, 1998

LaBuda CJ, Fuchs PN: The anxiolytic effect of acute ethanol or diazepam exposure is unaltered in mu-opioid receptor knockout mice. Brain Res Bull 55:755–760, 2001

Ledent C, Vaugeois JM, Schiffmann SN, et al: Aggressiveness, hypoalgesia and high blood pressure in mice lacking the adenosine A2a receptor. Nature 388:674–678, 1997

LeDoux JE: Emotion circuits in the brain. Annu Rev Neurosci 23:155–184, 2000

Lenox RH, Hahn CG: Overview of the mechanism of action of lithium in the brain: fifty-year update. J Clin Psychiatry 61 (suppl 9):5–15, 2000

Lenox RH, Watson DG, Patel J, et al: Chronic lithium administration alters a prominent PKC substrate in rat hippocampus. Brain Res 570(1–2):333–340, 1992

Lenox RH, Gould TD, Manji HK: Endophenotypes in bipolar disorder. Am J Med Genet 114(4):391–406, 2002

Levine J, Barak Y, Gonzalves M, et al: Double-blind, controlled trial of inositol treatment of depression. Am J Psychiatry 152(5):792–794, 1995

Levine J, Panchalingram K, Rapoport A, et al: Increased cerebrospinal fluid glutamine levels in depressed patients. Biol Psychiatry 47(7): 586–593, 2000

Li Q, Wichems C, Heils A, et al: Reduction in the density and expression, but not G-protein coupling, of serotonin receptors (5-$HT_{1A}$) in 5-HT transporter knock-out mice: gender and brain region differences. J Neurosci 20:7888–7895, 2000

Li X, Tizzano JP, Griffey K, et al: Antidepressant-like actions of an AMPA receptor potentiator (LY392098). Neuropharmacology 40(8):1028–1033, 2001

Li X, Bijur GN, Jope RS: Glycogen synthase kinase 3-beta, mood stabilizers, and neuroprotection. Bipolar Disord 4:137–144, 2002

Linden AM, Vaisanen J, Lakso M, et al: Expression of neurotrophins BDNF and NT-3, and their receptors in rat brain after administration of antipsychotic and psychotrophic agents. J Mol Neurosci 14:27–37, 2000

Lopez JF, Chalmers DT, Little KY, et al: A.E. Bennett Research Award. Regulation of serotonin$_{1A}$, glucocorticoid, and mineralocorticoid receptor in rat and human hippocampus: implications for the neurobiology of depression. Biol Psychiatry 43(8):547–573, 1998

Lyons DM: Stress, depression, and inherited variation in primate hippocampal and prefrontal brain development. Psychopharmacol Bull 36(1):27–43, 2002

Lyons DM, Yang C, Sawyer-Glover AM, et al: Early life stress and inherited variation in monkey hippocampal volumes. Arch Gen Psychiatry 58(12):1145–1151, 2001

Maccarrone M, Valverde O, Barbaccia ML, et al: Age-related changes of anandamide metabolism in CB1 cannabinoid receptor knockout mice: correlation with behaviour. Eur J Neurosci 15:1178–1186, 2002

MacQueen GM, Ramakrishnan K, Croll SD, et al: Performance of heterozygous brain-derived neurotrophic factor knockout mice on behavioral analogues of anxiety, nociception, and depression. Behav Neurosci 115:1145–1153, 2001

Maldonado R, Smadja C, Mazzucchelli C, et al: Altered emotional and locomotor responses in mice deficient in the transcription factor CREM. Proc Natl Acad Sci U S A 96:14094–14099, 1999

Mamounas LA, Blue ME, Siuciak JA, et al: Brain-derived neurotrophic factor promotes the survival and sprouting of serotonergic axons in rat brain. J Neurosci 15(12):7929–7939, 1995

Manev H, Uz T, Smalheiser NR, et al: Antidepressants alter cell proliferation in the adult brain in vivo and in neural cultures in vitro. Eur J Pharmacol 411(1–2):67–70, 2001

Manji HK: G proteins: implications for psychiatry. Am J Psychiatry 149(6): 746–760, 1992

Manji HK, Chen G: Post-receptor signaling pathways in the pathophysiology and treatment of mood disorders. Curr Psychiatry Rep 2(6): 479–489, 2000

Manji HK, Chen G: PKC, MAP kinases and the bcl-2 family of proteins as long-term targets for mood stabilizers. Mol Psychiatry 7 (suppl 1): S46–S56, 2002

Manji HK, Duman RS: Impairments of neuroplasticity and cellular resilience in severe mood disorder: implications for the development of novel therapeutics. Psychopharmacol Bull 35(2):5–49, 2001

Manji HK, Lenox RH: Ziskind-Somerfeld Research Award. Protein kinase C signaling in the brain: molecular transduction of mood stabilization in the treatment of manic-depressive illness. Biol Psychiatry 46(10):1328–1351, 1999

Manji HK, Lenox RH: Signaling: cellular insights into the pathophysiology of bipolar disorder. Biol Psychiatry 48(6):518–530, 2000

Manji HK, Etcheberrigaray R, Chen G, et al: Lithium decreases membrane-associated protein kinase C in hippocampus: selectivity for the alpha isozyme. J Neurochem 61(6):2303–2310, 1993

Manji HK, Moore GJ, Chen G: Lithium at 50: have the neuroprotective effects of this unique cation been overlooked? Biol Psychiatry 46(7): 929–940, 1999

Manji HK, Drevets WC, Charney DS: The cellular neurobiology of depression. Nat Med 7(5):541–547, 2001a

Manji HK, Moore GJ, Chen G: Bipolar disorder: leads from the molecular and cellular mechanisms of action of mood stabilizers. Br J Psychiatry Suppl 41:S107–S119, 2001b

Manji HK, Quiroz JA, Sporn J, et al: Enhancing synaptic plasticity and cellular resilience to develop novel, improved therapeutics. Biol Psychiatry (in press)

Martin M, Ledent C, Parmentier M, et al: Involvement of CB1 cannabinoid receptors in emotional behaviour. Psychopharmacology (Berl) 159:379–387, 2002

Martinek S, Inonog S, Manoukian AS, et al: A role for the segment polarity gene shaggy/GSK-3 in the Drosophila circadian clock. Cell 105(6):769–779, 2001

Mathis P, Schmitt L, Benatia M, et al: Plasma amino acid disturbances and depression [in French]. Encephale 14(2):77–82, 1988

Matsumoto T, Numakawa T, Adachi N, et al: Brain-derived neurotrophic factor enhances depolarization-evoked glutamate release in cultured cortical neurons. J Neurochem 79(3):522–530, 2001

McCullumsmith RE, Meador-Woodruff JH: Striatal excitatory amino acid transporter transcript expression in schizophrenia, bipolar disorder, and major depressive disorder. Neuropsychopharmacology 26(3):368–375, 2002

McEwen BS: Stress and hippocampal plasticity. Annu Rev Neurosci 22: 105–122, 1999

McQuade R, Young AH: Future therapeutic targets in mood disorders: the glucocorticoid receptor. Br J Psychiatry 177:390–395, 2000

Meador-Woodruff JH, Hogg AJ Jr, Smith RE: Striatal ionotropic glutamate receptor expression in schizophrenia, bipolar disorder, and major depressive disorder. Brain Bull Res 55(5):631–640, 2001

Michael-Titus AT, Bains S, Jeetle J, et al: Imipramine and phenelzine decrease glutamate overflow in the prefrontal cortex—a possible mechanism of neuroprotection in major depression? Neuroscience 100(4): 681–684, 2000

Miguel-Hidalgo JJ, Rajkowska G: Morphological brain changes in depression: can antidepressants reverse them? CNS Drugs 16(6):361–372, 2002

Milad MR, Quirk GJ: Neurons in medial prefrontal cortex signal memory for fear extinction. Nature 420(6911):70–74, 2002

Milligan G, Wakelam M: G Proteins: Signal Transduction and Disease. San Diego, CA, Academic Press, 1992

Mineur YS, Sluyter F, de WS, et al: Behavioral and neuroanatomical characterization of the Fmr1 knockout mouse. Hippocampus 12:39–46, 2002

Miyakawa T, Yared E, Pak JH, et al: Neurogranin null mutant mice display performance deficits on spatial learning tasks with anxiety related components. Hippocampus 11:763–775, 2001

Mohler H, Fritschy JM, Rudolph U: A new benzodiazepine pharmacology. J Pharmacol Exp Ther 300(1):2–8, 2002

Moore GJ, Bebchuk JM, Hasanat K, et al: Lithium increases N-acetyl-aspartate in the human brain: in vivo evidence in support of bcl-2's neurotrophic effects? Biol Psychiatry 48(1):1–8, 2000a

Moore GJ, Bebchuk JM, Wilds IB, et al: Lithium-induced increase in human brain grey matter. Lancet 356(9237):1241–1242, 2000b

Moore G, Rajarethinam R, Cortese B, et al: Regionally specific increases in human brain gray matter with chronic lithium treatment, in Abstracts of the Society for Neuroscience, Vol 27, Program No 111.8, 2001

Murray CJ, Lopez AD: Global mortality, disability, and the contribution of risk factors: Global Burden of Disease Study. Lancet 349(9063): 1436–1442, 1997

Nakamura E, Kadomatsu K, Yuasa S, et al: Disruption of the midkine gene (Mdk) resulted in altered expression of a calcium binding protein in the hippocampus of infant mice and their abnormal behaviour. Genes Cells 3:811–822, 1998

Nakamura S: Antidepressants induce regeneration of catecholaminergic axon terminals in the rat cerebral cortex. Neurosci Lett 111(1–2): 64–68, 1990

Nemeroff CB: Recent advances in the neurobiology of depression. Psychopharmacol Bull 36 (suppl 2):6–23, 2002
Nemeroff CB, Owens MJ: Treatment of mood disorders. Nat Neurosci 5(suppl):1068–1070, 2002
Nestler EJ: Antidepressant treatments in the 21st century. Biol Psychiatry 44(7):526–533, 1998
Nestler EJ, Barrot M, DiLeone RJ, et al: Neurobiology of depression. Neuron 34(1):13–25, 2002
Nibuya M, Morinobu S, Duman RS: Regulation of BDNF and trkB mRNA in rat brain by chronic electroconvulsive seizure and antidepressant drug treatments. J Neurosci 15(11):7539–7547, 1995
Nibuya M, Nestler EJ, Duman RS: Chronic antidepressant administration increases the expression of cAMP response element binding protein (CREB) in rat hippocampus. J Neurosci 16(7):2365–2372, 1996
Nibuya M, Takahashi M, Russell DS, et al: Repeated stress increases catalytic TrkB mRNA in rat hippocampus. Neurosci Lett 267(2):81–84, 1999
Nielsen DM, Derber WJ, McClellan DA, et al: Alterations in the auditory startle response in Fmr1 targeted mutant mouse models of fragile X syndrome. Brain Res 927:8–17, 2002
O'Donnell JM: Antidepressant-like effects of rolipram and other inhibitors of cyclic adenosine monophosphate phosphodiesterase on behavior maintained by differential reinforcement of low response rate. J Pharmacol Exp Ther 264(3):1168–1178, 1993
Oh YJ, Swarzenski BC, O'Malley KL: Overexpression of Bcl-2 in a murine dopaminergic neuronal cell line leads to neurite outgrowth. Neurosci Lett 202(3):161–164, 1996
Ongur D, Drevets WC, Price JL: Glial reduction in the subgenual prefrontal cortex in mood disorders. Proc Natl Acad Sci U S A 95(22): 13290–13295, 1998
Ontiveros A, Fontaine R, Breton G, et al: Correlation of severity of panic disorder and neuroanatomical changes on magnetic resonance imaging. J Neuropsychiatry Clin Neurosci 1(4):404–408, 1989
Otto C, Martin M, Wolfer DP, et al: Altered emotional behavior in PACAP-type-I-receptor-deficient mice. Brain Res Mol Brain Res 92: 78–84, 2001
Owens MJ, Nemeroff CB: Corticotropin-releasing factor antagonists in affective disorders. Expert Opin Investig Drugs 8(11):1849–1858, 1999
Palatnik A, Frolov K, Fux M, et al: Double-blind, controlled, crossover trial of inositol versus fluvoxamine for the treatment of panic disorder. J Clin Psychopharmacol 21(3):335–339, 2001

Pande AC, Greiner M, Adams JB, et al: Placebo-controlled trial of the CCK-B antagonist, CI-988, in panic disorder. Biol Psychiatry 46(6): 860–862, 1999

Pandey GN, Dwivedi Y, Pandey SC, et al: Protein kinase C in the postmortem brain of teenage suicide victims. Neurosci Lett 228(2):111–114, 1997

Pap M, Cooper GM: Role of glycogen synthase kinase–3 in the phosphatidylinositol 3–kinase/Akt cell survival pathway. J Biol Chem 273(32): 19929–19932, 1998

Parks CL, Robinson PS, Sibille E, et al: Increased anxiety of mice lacking the serotonin$_{1A}$ receptor. Proc Natl Acad Sci U S A 95:10734–10739, 1998

Parsons LH, Kerr TM, Tecott LH: 5-HT$_{1A}$ receptor mutant mice exhibit enhanced tonic, stress-induced and fluoxetine-induced serotonergic neurotransmission. J Neurochem 77: 607–617, 2001

Patapoutian A, Reichardt LF: Trk receptors: mediators of neurotrophin action. Curr Opin Neurobiol 11(3):272–280, 2001

Patel PD, Lopez JF, Lyons DM, et al: Glucocorticoid and mineralocorticoid receptor mRNA expression in squirrel monkey brain. J Psychiatr Res 34(6):383–392, 2000

Pattij T, Groenink L, Oosting RS, et al: GABA$_A$-benzodiazepine receptor complex sensitivity in 5-HT$_{1A}$ receptor knockout mice on a 129/Sv background. Eur J Pharmacol 447:67–74, 2002

Payne JL, Quiroz JA, Zarate CA, et al: Timing is everything: does the robust upregulation of noradrenergically regulated plasticity genes underlie the rapid antidepressant effects of sleep deprivation? Biol Psychiatry 52:921–926, 2002

Petronis A: Human morbid genetics revisited: relevance of epigenetics. Trends Genet 17(3):142–146, 2001

Phiel CJ, Klein PS: Molecular targets of lithium action. Annu Rev Pharmacol Toxicol 41:789–813, 2001

Phiel CJ, Zhang F, Huang EY, et al: Histone deacetylase is a direct target of valproic acid, a potent anticonvulsant, mood stabilizer, and teratogen. J Biol Chem 276(39):36734–36741, 2001

Poo MM: Neurotrophins as synaptic modulators. Nat Rev Neurosci 2(1): 24–32, 2001

Popova NK, Skrinskaya YA, Amstislavskaya TG, et al: Behavioral characteristics of mice with genetic knockout of monoamine oxidase type A. Neurosci Behav Physiol 31:597–602, 2001

Quiroz JA, Manji HK: Enhancing synaptic plasticity and cellular resilience to develop novel, improved treatments for mood disorders. Dialogues in Clinical Neuroscience 4:73–92, 2002

Raber J, Akana SF, Bhatnagar S, et al: Hypothalamic-pituitary-adrenal dysfunction in Apoe(–/–) mice: possible role in behavioral and metabolic alterations. J Neurosci 20:2064–2071, 2000

Ragnauth A, Schuller A, Morgan M, et al: Female preproenkephalin-knockout mice display altered emotional responses. Proc Natl Acad Sci U S A 98:1958–1963, 2001

Rajkowska G: Histopathology of the prefrontal cortex in major depression: what does it tell us about dysfunctional monoaminergic circuits? Prog Brain Res 126:397–412, 2000a

Rajkowska G: Postmortem studies in mood disorders indicate altered numbers of neurons and glial cells. Biol Psychiatry 48(8):766–477, 2000b

Rajkowska G: Cell pathology in bipolar disorder. Bipolar Disord 4(2): 105–116, 2002

Rajkowska G, Miguel-Hidalgo JJ, Wei J, et al: Morphometric evidence for neuronal and glial prefrontal cell pathology in major depression. Biol Psychiatry 45(9):1085–1098, 1999

Ramboz S, Oosting R, Amara DA, et al: Serotonin receptor 1A knockout: an animal model of anxiety-related disorder. Proc Natl Acad Sci U S A 95:14476–14481, 1998

Reul JM, Holsboer F: Corticotropin-releasing factor receptors 1 and 2 in anxiety and depression. Curr Opin Pharmacol 2(1):23–33, 2002

Reus VI, Wolkowitz OM: Antiglucocorticoid drugs in the treatment of depression. Expert Opin Investig Drugs 10(10):1789–1796, 2001

Riccio A, Ahn S, Davenport CM, et al: Mediation by a CREB family transcription factor of NGF-dependent survival of sympathetic neurons. Science 286(5448):2358–2361, 1999

Rodriguez G, Cogorno P, Gris A, et al: Regional cerebral blood flow and anxiety: a correlation study in neurologically normal patients. J Cereb Blood Flow Metab 9(3):410–416, 1989

Rupniak NM, Carlson EJ, Webb JK, et al: Comparison of the phenotype of NK1R–/– mice with pharmacological blockade of the substance P (NK1 ) receptor in assays for antidepressant and anxiolytic drugs. Behav Pharmacol 12:497–508, 2001

Ryves WJ, Harwood AJ: Lithium inhibits glycogen synthase kinase–3 by competition for magnesium. Biochem Biophys Res Commun 280(3): 720–725, 2001

Sanchez MM, Young LJ, Plotsky PM, et al: Distribution of corticosteroid receptors in the rhesus brain: relative absence of glucocorticoid receptors in the hippocampal formation. J Neurosci 20(12):4657–4668, 2000

Sandford JJ, Argyropoulos SV, Nutt DJ: The psychobiology of anxiolytic drugs, part 1: basic neurobiology. Pharmacol Ther 88(3):197–212, 2000

Santarelli L, Gobbi G, Debs PC, et al: Genetic and pharmacological disruption of neurokinin 1 receptor function decreases anxiety-related behaviors and increases serotonergic function. Proc Natl Acad Sci U S A 98:1912–1917, 2001

Sapolsky RM: Stress, glucocorticoids, and damage to the nervous system: the current state of confusion. Stress 1(1):1–19, 1996

Sapolsky RM: Glucocorticoids and hippocampal atrophy in neuropsychiatric disorders. Arch Gen Psychiatry 57(10):925–35, 2000

Sasaki K, Fan LW, Tien LT, et al: The interaction of morphine and gamma-aminobutyric acid (GABA)ergic systems in anxiolytic behavior: using mu-opioid receptor knockout mice. Brain Res Bull 57:689–694, 2002

Schaefer ML, Wong ST, Wozniak DF, et al: Altered stress-induced anxiety in adenylyl cyclase type VIII–deficient mice. J Neurosci 20:4809–4820, 2000

Schinder AF, Berninger B, Poo M: Postsynaptic target specificity of neurotrophin-induced presynaptic potentiation. Neuron 25(1):151–163, 2000

Schramm NL, McDonald MP, Limbird LE: The alpha$_{2a}$-adrenergic receptor plays a protective role in mouse behavioral models of depression and anxiety. J Neurosci 21:4875–4882, 2001

Scott AI, Perini AF, Shering PA, et al: In-patient major depression: is rolipram as effective as amitriptyline? Eur J Clin Pharmacol 40(2):127–129, 1991

Selemon LD, Kleinman JE, Herman MM, et al: Smaller frontal gray matter volume in postmortem schizophrenic brains. Am J Psychiatry 159(12): 1983–1991, 2002

Sheline YI, Sanghavi M, Mintun MA, et al: Depression duration but not age predicts hippocampal volume loss in medically healthy women with recurrent major depression. J Neurosci 19(12):5034–5043, 1999

Sibille E, Pavlides C, Benke D, et al: Genetic inactivation of the serotonin$_{1A}$ receptor in mice results in downregulation of major GABA$_A$ receptor alpha subunits, reduction of GABA$_A$ receptor binding, and benzodiazepine-resistant anxiety. J Neurosci 20:2758–2765, 2000

Sillaber I, Rammes G, Zimmermann S, et al: Enhanced and delayed stress-induced alcohol drinking in mice lacking functional CRH1 receptors. Science 296:931–933, 2002

Skolnick P: Antidepressants for the new millennium. Eur J Pharmacol 375(1–3):31–40, 1999

Smith MA, Makino S, Kvetnansky R, et al: Stress and glucocorticoids affect the expression of brain-derived neurotrophic factor and neurotrophin-3 mRNAs in the hippocampus. J Neurosci 15 (3, pt 1):1768–1777, 1995

Spiegel A: G Proteins, Receptors, and Disease. Totowa, NJ., Humana Press, 1998

Sporn J, Sachs G: The anticonvulsant lamotrigine in treatment-resistant manic-depressive illness. J Clin Psychopharmacol 17(3):185–189, 1997

Steffens DC, Krishnan KR: Structural neuroimaging and mood disorders: recent findings, implications for classification, and future directions. Biol Psychiatry 43(10):705–712, 1998

Steffens DC, Helms MJ, Krishnan KR, et al: Cerebrovascular disease and depression symptoms in the Cardiovascular Health Study. Stroke 30(10):2159–2166, 1999

Stein MB, Koverola C, Hanna C, et al: Hippocampal volume in women victimized by childhood sexual abuse. Psychol Med 27(4):951–959, 1997

Steiner H, Fuchs S, Accili D: D3 dopamine receptor–deficient mouse: evidence for reduced anxiety. Physiol Behav 63:137–141, 1997

Stewart RS, Devous MD Sr, Rush AJ, et al: Cerebral blood flow changes during sodium-lactate–induced panic attacks. Am J Psychiatry 145(4): 442–449, 1988

Stoll AL, Renshaw PF, Yurgelun-Todd DA, et al: Neuroimaging in bipolar disorder: what have we learned? Biol Psychiatry 48(6):505–517, 2000

Stork O, Welzl H, Wotjak CT, et al: Anxiety and increased 5-$HT_{1A}$ receptor response in NCAM null mutant mice. J Neurobiol 40:343–355, 1999

Stork O, Ji FY, Kaneko K, et al: Postnatal development of a GABA deficit and disturbance of neural functions in mice lacking GAD65. Brain Res 865:45–58, 2000a

Stork O, Welzl H, Wolfer D, et al: Recovery of emotional behaviour in neural cell adhesion molecule (NCAM) null mutant mice through transgenic expression of NCAM180. Eur J Neurosci 12:3291–3306, 2000b

Stout SC, Owens MJ, Nemeroff CB: Neurokinin(1) receptor antagonists as potential antidepressants. Annu Rev Pharmacol Toxicol 41:877–906, 2001

Strakowski SM, Wilson DR, Tohen M, et al: Structural brain abnormalities in first-episode mania. Biol Psychiatry 33(8–9):602–609, 1993

Strakowski SM, Adler CM, DelBello MP: Volumetric MRI studies of mood disorders: do they distinguish unipolar and bipolar disorder? Bipolar Disord 4(2):80–88, 2002

Szabo ST, Blier P: Effects of serotonin (5-hydroxytryptamine, 5-HT) reuptake inhibition plus 5-HT(2A) receptor antagonism on the firing activity of norepinephrine neurons. J Pharmacol Exp Ther 302(3):983–991, 2002

Takahashi M, Terwilliger R, Lane C, et al: Chronic antidepressant administration increases the expression of cAMP-specific phosphodiesterase 4A and 4B isoforms. J Neurosci 19(2):610–618, 1999

Takashima A, Noguchi K, Sato K, et al: Tau protein kinase I is essential for amyloid beta-protein–induced neurotoxicity. Proc Natl Acad Sci U S A 90(16):7789–7793, 1993

Taylor WD, Payne ME, Krishnan KR, et al: Evidence of white matter tract disruption in MRI hyperintensities. Biol Psychiatry 50(3):179–183, 2001

Timpl P, Spanagel R, Sillaber I, et al: Impaired stress response and reduced anxiety in mice lacking a functional corticotropin-releasing hormone receptor. (see comments). Nat Genet 19:162–166, 1998

Tronche F, Kellendonk C, Kretz O, et al: Disruption of the glucocorticoid receptor gene in the nervous system results in reduced anxiety. Nat Genet 23:99–103, 1999

Tsai G, Coyle JT: N-Acetylaspartate in neuropsychiatric disorders. Prog Neurobiol 46(5):531–540, 1995

Uhde TW, Boulenger JP, Post RM, et al: Fear and anxiety: relationship to noradrenergic function. Psychopathology 17 (suppl 3):8–23, 1984

Ullian EM, Sapperstein SK, Christopherson KS, et al: Control of synapse number by glia. Science 291(5504):657–661, 2001

van Praag H, Kempermann G, Gage FH: Running increases cell proliferation and neurogenesis in the adult mouse dentate gyrus. Nat Neurosci 2(3):266–270, 1999

Vincent SL, Todtenkopf MS, Benes FM: A comparison of the density of pyramidal and nonpyramidal neurons in the anterior cingulate cortex in schizophrenics and manic depressives. Soc Neurosci 23:2199, 1997

Wachtel H, Schneider HH: Rolipram, a novel antidepressant drug, reverses the hypothermia and hypokinesia of monoamine-depleted mice by an action beyond postsynaptic monoamine receptors. Neuropharmacology 25(10):1119–1126, 1986

Walther T, Balschun D, Voigt JP, et al: Sustained long term potentiation and anxiety in mice lacking the Mas protooncogene. J Biol Chem 273: 11867–11873, 1998

Walther T, Voigt JP, Fink H, et al: Sex specific behavioural alterations in Mas-deficient mice. Behav Brain Res 107:105–109, 2000

Wang HY, Friedman E: Enhanced protein kinase C activity and translocation in bipolar affective disorder brains. Biol Psychiatry 40:568–575, 1996

Watanabe Y, Gould E, Daniels DC, et al: Tianeptine attenuates stress-induced morphological changes in the hippocampus. Eur J Pharmacol 222(1):157–162, 1992

Watson DG, Watterson JM, Lenox RH: Sodium valproate down-regulates the myristoylated alanine-rich C kinase substrate (MARCKS) in immortalized hippocampal cells: a property of protein kinase C–mediated mood stabilizers. J Pharmacol Exp Ther 285(1):307–316, 1998

Webster MJ, Knable MB, O'Grady J, et al: Regional specificity of brain glucocorticoid receptor mRNA alterations in subjects with schizophrenia and mood disorders. Mol Psychiatry 7(9):985–994, 1999

Weinstein LS: The role of tissue-specific imprinting as a source of phenotypic heterogeneity in human disease. Biol Psychiatry 50(12): 927–931, 2001

Weintraub B: Molecular Endocrinology: Basic Concepts and Clinical Correlations. New York, NY, Raven, 1995

Weng G, Bhalla US, Iyengar R: Complexity in biological signaling systems. Science 284(5411):92–96, 1999

Woods SW, Koster K, Krystal JK, et al: Yohimbine alters regional cerebral blood flow in panic disorder. Lancet 2(8612):678, 1988

Yamada K, Santo-Yamada Y, Wada K: Restraint stress impaired maternal behavior in female mice lacking the neuromedin B receptor (NMB-R) gene. Neurosci Lett 330: 163-6., 2002a

Yamada K, Santo-Yamada Y, Wada E, et al: Role of bombesin (BN)–like peptides/receptors in emotional behavior by comparison of three strains of BN-like peptide receptor knockout mice. Mol Psychiatry 7: 113–117, 2002b

Yamada K, Wada E, Yamano M, et al: Decreased marble burying behavior in female mice lacking neuromedin-B receptor (NMB-R) implies the involvement of NMB/NMB-R in 5-HT neuron function. Brain Res 942:71–78, 2002c

Young AH: The effects of glucocorticoid antagonists in unipolar and bipolar disorder. Presentation at the Society of Biological Psychiatry 57th Annual Scientific Convention, Philadelphia, PA, May 2002

Yuan PX, Huang LD, Jiang YM, et al: The mood stabilizer valproic acid activates mitogen-activated protein kinases and promotes neurite growth. J Biol Chem 276(34):31674–31683, 2001

Zarate CAJ, Quiroz JA, Payne JL, et al: Modulators of the glutamatergic system: implications for the development of improved therapeutics in mood disorders. Psychopharmacol Bull (in press)

Zeller E, Stief HJ, Pflug B, et al: Results of a phase II study of the antidepressant effect of rolipram. Pharmacopsychiatry 17(6):188–190, 1984

Zobel AW, Nickel T, Kunzel HE, et al: Effects of the high-affinity corticotropin-releasing hormone receptor 1 antagonist R121919 in major depression: the first 20 patients treated. J Psychiatr Res 34(3):171–181, 2000

**Appendix 1:** Knockout mouse studies that have been used to test for anxiety-like behaviors

| Study | Knockout | Effect on anxiety behavior | Comments/additional phenotypes |
|---|---|---|---|
| Li et al. 2000 | 5-HT transporter | Not tested | Reduced 5-$HT_{1A}$, $G_o$, and $G_i1$ densities |
| Gross et al. 2002 | 5-$HT_{1A}$ receptor | Decreased anxiety | Anxiogenic phenotype rescued by postnatal hippocampal/cortical, but not adult or raphe, expression |
| He et al. 2001 | 5-$HT_{1A}$ receptor | Not tested | Knockout displayed much larger increase in 5-HT following acute fluoxetine administration compared with wild type |
| Heisler et al. 1998 | 5-$HT_{1A}$ receptor | Increased anxiety | Decreased depressive-like and exploratory behavior (tail suspension immobility and novelty seeking) |
| Parks et al. 1998 | 5-$HT_{1A}$ receptor | Increased anxiety (greater in males) | Decreased depressive-like behavior (forced swim) |
| Parsons et al. 2001 | 5-$HT_{1A}$ receptor | Not tested | Greatly increased frontal cortex 5-HT in mutants in response to stressors, an effect potentiated by acute fluoxetine administration |

**Appendix 1:** Knockout mouse studies that have been used to test for anxiety-like behaviors *(continued)*

| Study | Knockout | Effect on anxiety behavior | Comments/additional phenotypes |
|---|---|---|---|
| Ramboz et al. 1998 | 5-$HT_{1A}$ receptor | Increased anxiety (greater in males) | Decreased depressive-like behavior (forced swim) |
| Sibille et al. 2000 | 5-$HT_{1A}$ receptor | Data not shown | Anxiolytic and sedative properties of BZDs blocked by mutation; reduced $GABA_A$ $\alpha$ subunit densities in amygdala and parietal cortex |
| Bouwknecht et al. 2001 | 5-$HT_{1B}$ receptor | Not tested | Impulse disinhibition: increased aggression, social interaction, locomotion, etc. |
| Brunner et al. 1999 | 5-$HT_{1B}$ receptor | Decreased | Hyperlocomotion |
| Johansson et al. 2001 | Adenosine $A_1$ receptor | Increased anxiety | Hyperalgesia |
| Ledent et al. 1997 | Adenosine $A_{2a}$ receptor | Increased anxiety | Increased aggression; hypoalgesia |
| Schaefer et al. 2000 | Adenylate cyclase, type VIII | Decreased anxiety | Basal levels of corticosterone increased by chronic stress in knockout compared with wild type |
| Schramm et al. 2001 | $\alpha_2$ adrenergic receptor | Increased anxiety | Depressive-like behavior in forced swim test; insensitive to imipramine |

**Appendix 1:** Knockout mouse studies that have been used to test for anxiety-like behaviors *(continued)*

| Study | Knockout | Effect on anxiety behavior | Comments/additional phenotypes |
|---|---|---|---|
| Grootendorst et al. 2001 | ApoE | No anxiety-like effect in light-dark test | Elevated basal corticosterone, but reduced corticosterone 5 days after severe stress |
| Raber et al. 2000 | ApoE | Increased anxiety | Increased corticosterone in response to stress or ACTH in older mice |
| MacQueen et al. 2001 | BDNF | No effect | Depressive-like behavior (?); no effect in forced swim, but enhanced learned helplessness |
| Yamada et al. 2002b | Bombesin receptor, subtype 3 | Unclear | Increased open-arm time, but increased risk-assessment behavior, in elevated plus maze |
| S. Chen et al. 1994 | CaMKIIα | Reduced | Increased aggression, decreased mating |
| Maccarrone et al. 2002 | CB1 receptor | Increased anxiety | May be age-dependent, with reduced anxiety in older mice |
| Martin et al. 2002 | CB1 receptor | Increased anxiety | Increased aggression; depressive-like behavior |

**Appendix 1:** Knockout mouse studies that have been used to test for anxiety-like behaviors *(continued)*

| Study | Knockout | Effect on anxiety behavior | Comments/additional phenotypes |
|---|---|---|---|
| Gogos et al. 1998 | COMT | Increased in females | Increased aggression in male heterozygotes; sexually dimorphic effects on monoamine concentrations |
| Maldonado et al. 1999 | CREM | Decreased anxiety | Increased locomotion; possible alterations in circadian rhythms |
| Sillaber et al. 2002 | CRFR1 | Not tested | Increased voluntary ethanol consumption following social stress; upregulation of NMDA receptor 2b |
| Contarino et al. 1999 | CRFR1 | Decreased anxiety | Reduced exploratory behavior; cognitive impairment? |
| Contarino et al. 2000 | CRFR1 | Blunted reponse to CRH anxiogenesis | No effect on CRH anorexia |
| Timpl et al. 1998 | CRFR1 | Decreased anxiety | Increased locomotion and exploratory activity and reduced corticosterone concentrations; similar phenotype following ethanol withdrawal |

**Appendix 1:** Knockout mouse studies that have been used to test for anxiety-like behaviors *(continued)*

| Study | Knockout | Effect on anxiety behavior | Comments/additional phenotypes |
|---|---|---|---|
| Bale et al. 2002 | CRFR1/CRFR2 double | Decreased in females, increased in males | Maternal genotype suggests importance of rearing behavior |
| Kishimoto et al. 2000 | CRFR2 | Increased anxiety in males | Decreased pCREB in several brain regions |
| Steiner et al. 1997 | $D_3$ | Decreased anxiety | No effect on locomotion |
| Dulawa et al. 1999 | $D_4$ | Slight increase | Prominent reductions in exploratory behavior |
| Falzone et al. 2002 | $D_4$ | Increased anxiety | Increased anxiety blocked by BZD or ethanol; no genotype effect on conditioned fear acquisition. |
| Filliol et al. 2000 | δ opioid receptor | Increased | Increased depressive-like behavior (forced swim) |
| Krezel et al. 2001 | Estrogen α | No effect | Nonsignificant trend toward reduced anxiety |
| Krezel et al. 2001 | Estrogen β | Increased anxiety | Enhanced induction of LTP; increased $5\text{-}HT_{1A}$ mRNA in amygdala |
| Mineur et al. 2002 | Fmr1 | No effect | Impaired learning; hyperlocomotion |

**Appendix 1:** Knockout mouse studies that have been used to test for anxiety-like behaviors *(continued)*

| Study | Knockout | Effect on anxiety behavior | Comments/additional phenotypes |
|---|---|---|---|
| Nielsen et al. 2002 | Fmr1 | No effect | Increased startle to moderate, but decreased startle to loud acoustic stimuli |
| Kash et al. 1999 | $GAD_{65}$ | Increased anxiety | Reduced sensitivity to BZDs |
| Stork et al. 2000a | $GAD_{65}$ | Increased anxiety | Decreased forced-swim immobility; reduced aggression; developmental deficit in GABA levels coinciding with seizures and increased mortality |
| Blednov et al. 2001a | GIRK2 | Decreased anxiety | Hyperactivity; perhaps a functional block of $D_3$ receptors |
| Blednov et al. 2001b | GIRK2 | Decreased anxiety | Failure to exhibit anxiolysis or hyperactivity in response to ethanol |
| Tronche et al. 1999 | Glucocorticoid receptor | Decreased anxiety | Elevated corticosterone, reduced ACTH levels |
| Kustova et al. 1999 | Interferon-γ | Increased anxiety | Strain difference (C57B1/6 > BALB/c) |
| Cases et al. 1995 | MAO-A | Mixed; increased fearfulness in pups, more center field time in adults | Increased aggression and forced-swim struggle time |

**Appendix 1:** Knockout mouse studies that have been used to test for anxiety-like behaviors *(continued)*

| Study | Knockout | Effect on anxiety behavior | Comments/additional phenotypes |
|---|---|---|---|
| Popova et al. 2001 | MAO-A | No effect | Increased male aggression; reduced startle response |
| Nakamura et al. 1998 | Midkine | Increased anxiety at 4 weeks but not 8 weeks | Working memory deficit |
| Filliol et al. 2000 | μ opioid receptor | Decreased | Reduced depressive-like behavior (forced swim) |
| LaBuda and Fuchs 2001 | μ opioid receptor | No effect | Increased motor activity; anxiolytic effects of ethanol and BZDs unaltered |
| Sasaki et al. 2002 | μ opioid receptor | Decreased anxiety | Blocked sensitivity to $GABA_A$ agonist as a result of the mutation , yet increased $GABA_A$ density |
| Stork et al. 1999, 2000b | NCAM | Increased anxiety | Increased aggression; decreased forced-swim immobility; hypersensitivity to $5\text{-}HT_{1A}$ agonist |
| Miyakawa et al. 2001 | Neurogranin | Increased anxiety | Spatial memory deficits in homo-zygotes (anxiogenic phenotype was similar in homo- and heterozygotes) |

**Appendix 1:** Knockout mouse studies that have been used to test for anxiety-like behaviors *(continued)*

| **Study** | **Knockout** | **Effect on anxiety behavior** | **Comments/additional phenotypes** |
|---|---|---|---|
| Rupniak et al. 2001 | Neurokinin-1 receptor (substance P receptor) | Possible decrease in anxiety | Reduced aggression in both knockout and wild type treated with a substance P antagonist; reduced depressive-like behavior on forced swim and tail suspension seen in knockout |
| Santarelli et al. 2001 | Neurokinin-1 receptor (substance P receptor) | Decreased anxiety | Desensitization of 5-$HT_{1A}$ autoreceptors; increased dorsal raphe 5-HT output |
| Koster et al. 1999 | Nociceptin | Increased anxiety | Elevated corticosterone; increased basal pain threshold |
| Otto et al. 2001 | PACAP type I receptor | Decreased anxiety | Increased locomotion; anxiogenic and hyperactive phenotypes not observed in a forebrain-specific deletion |
| Ikegaya et al. 2001 | PIMT | Decreased anxiety | Quite likely the result of more general cognitive impairments |

**Appendix 1:** Knockout mouse studies that have been used to test for anxiety-like behaviors *(continued)*

| Study | Knockout | Effect on anxiety behavior | Comments/additional phenotypes |
|---|---|---|---|
| Hodge et al. 1999 | PKC ε | | Greater behavioral sensitivity to alcohol and BZD agonists |
| Bowers et al. 2000 | PKC γ | Decreased | |
| Bowers et al. 2001 | PKC γ | | Decreased sensitivity to ethanol, increased sensitivity to BZDs |
| Ragnauth et al. 2001 | Preproenkephalin | Increased anxiety | Increased female sexual approach behavior in heterozygotes |
| Linden et al. 2002 | mGluR8 | Increased anxiety | Effects abolished by increasing anxiety-provoking stimuli of the experiment; wild type became more anxious, whereas knockout maintained already high levels of anxiety |
| Collinson et al. 2002 | $GABA_A$ receptor, $\alpha_5$ subunit | No effect on anxiety | Improved spatial learning; altered hippocampal electrophysiology |
| Pattij et al. 2002 | $5\text{-}HT_{1A}$ receptor | Increased anxiety (physiological/autonomic) | Stress-induced hyperthermia greater in knockout; sensitive to diazepam but not to flesinoxan |
| Walther et al. 1998, 2000 | *Mas* proto-oncogene | Increased anxiety in males | Enhanced durability of dentate gyrus LTP; no effect in females |

**Appendix 1:** Knockout mouse studies that have been used to test for anxiety-like behaviors *(continued)*

| Study | Knockout | Effect on anxiety behavior | Comments/additional phenotypes |
|---|---|---|---|
| Yamada 2002a, 2002b, 2002c | Neuromedin B receptor (bombesin-like peptide) | Increased anxiety | Greater disruption of maternal behaviors after 30 minutes of restraint stress; however, male mice display decreased risk-assessment behavior in elevated plus maze |

*Note.* ACTH = adrenocorticotropic hormone; ApoE = apolipoprotein E; BDNF = brain-derived neurotrophic factor; BZD = benzodiazepine; CaMKIIα = calcium/calmodulin–dependent kinase II α; COMT = catechol-*O*-methyltransferase; CREB = cAMP response element binding protein; CREM = cAMP response element modulating protein; CRFR1 = CRF receptor 1; CRFR2 = CRF receptor 2; CRH = corticotropin-releasing hormone; $D_3$ = $dopamine_3$ receptor; $D_4$ = $dopamine_4$ receptor; Fmr1 = fragile X mental retardation syndrome 1; GABA = γ-aminobutyric acid; $GABA_A$ = γ-aminobutyric $acid_A$ receptor; $GAD_{65}$ = glutamate decarboxylase, 65 kDa; $G_i1$ = guanine nucleotide binding protein, α subunit, inhibiting activity type 1; GIRK2 = G protein–activated, inwardly rectifying potassium channel 2; $G_o$ = guanine nucleotide binding protein, α subunit, olfactory type; 5-HT = serotonin; 5-$HT_{1A}$ = $serotonin_{1A}$; 5-$HT_{1B}$ = $serotonin_{1B}$; mGluR8 = metabotropic glutamate 8 receptor; LTP = long-term potentiation; MAO-A = monoamine oxidase A; NCAM = neural cell adhesion molecule; NMDA = *N*-methyl-D-aspartate; PACAP = pituitary adenylate cyclase activating polypeptide; pCREB = phospho-CREB; PIMT = protein L-isoaspartyl methyltransferase; PKC = protein kinase C.

**Appendix 2:** Potential targets for the development of new treatments for mood and anxiety disorders

| Molecule | Hypothesized involvement in mood or anxiety disorders or Rx | Function plausibly relevant to mood or anxiety disorders | Findings from animal models | Direct or surrogate human evidence | Observations from clinical treatment studies |
|---|---|---|---|---|---|
| $5\text{-HT}_{1A}$/ $5\text{-HT}_{1B}$ antagonists | $5\text{-HT}_{1A}$/$5\text{-HT}_{1B}$ antagonists may augment AD response | Somatodendritic $5\text{-HT}_{1A}$ receptors regulate 5-HT neuron firing; nerve terminal $5\text{-HT}_{1B}$ receptors facilitate 5-HT release; blockade increases 5-HT throughput via two mechanisms | Co-administration of $5\text{-HT}_{1A}$/ $5\text{-HT}_{1B}$ antagonists facilitates AD-induced 5-HT throughput | Both PET and neuroendocrine studies suggest reduced $5\text{-HT}_{1A}$ levels/function in depression; however, these studies have not investigated somatodendritic receptors | Equivocal results to date; pindolol may not be the ideal drug to test hypothesis |

**Appendix 2:** Potential targets for the development of new treatments for mood and anxiety disorders *(continued)*

| Molecule | Hypothesized involvement in mood or anxiety disorders or Rx | Function plausibly relevant to mood or anxiety disorders | Findings from animal models | Direct or surrogate human evidence | Observations from clinical treatment studies |
|---|---|---|---|---|---|
| $5\text{-}HT_2$ antagonists | $5\text{-}HT_2$ antagonists may possess AD and anxiolytic effects | $5\text{-}HT_2$ receptors are widely distributed in the CNS; may regulate DA throughput; important roles in regulating sleep and appetite | Many ADs downregulate $5\text{-}HT_2$ receptors (but not ECS) | Findings from postmortem brain and PET studies are inconclusive; however, PET studies suggest that ADs reduce $5\text{-}HT_2$ binding; elevated platelet $5\text{-}HT_2$ binding in depression | Agents with $5\text{-}HT_2$ antagonism (e.g., mirtazapine, clozapine, serazepine) have antidepressant and/or anxiolytic effects; no clinical studies with selective agents |
| $5\text{-}HT_3$ antagonists/ partial agonists | $5\text{-}HT_3$ antagonists may also be important in anxiolysis | | Microinjection of $5\text{-}HT_3$ antagonists into amygdala mimics behavioral effects of BZD microinjection | | Many effective anxiolytics display $5\text{-}HT_3$ affinity; more studies are needed on selective agents |

**Appendix 2:** Potential targets for the development of new treatments for mood and anxiety disorders *(continued)*

| Molecule | Hypothesized involvement in mood or anxiety disorders or Rx | Function plausibly relevant to mood or anxiety disorders | Findings from animal models | Direct or surrogate human evidence | Observations from clinical treatment studies |
|---|---|---|---|---|---|
| $5\text{-}HT_7$ antagonists | $5\text{-}HT_7$ antagonists may have AD effects and/or ameliorate sleep/circadian disturbances | $5\text{-}HT_7$ receptors are widely distributed in the CNS; may regulate circadian rhythms; positively linked to cAMP generation | ADs and clozapine appear to interact with and/or regulate the expression of $5\text{-}HT_7$ | None | None |

**Appendix 2:** Potential targets for the development of new treatments for mood and anxiety disorders *(continued)*

| **Molecule** | **Hypothesized involvement in mood or anxiety disorders or Rx** | **Function plausibly relevant to mood or anxiety disorders** | **Findings from animal models** | **Direct or surrogate human evidence** | **Observations from clinical treatment studies** |
|---|---|---|---|---|---|
| $\alpha_2$ antagonists administered during REM sleep | A very investigational strategy as a rapidly acting AD | $\alpha_2$ antagonists increase firing of LC and release of NE; enhancing NE function may have AD effects; activating central NE projections during REM sleep may allow released NE to interact with a primed, sensitized postsynaptic environment | Agents that enhance NE throughput have AD efficacy; sleep deprivation, which has a rapid AD effect, enhances the expression of "plasticity molecules" (e.g., CREB, BDNF) via a NE-dependent mechanism | Sleep deprivation exerts a rapid AD effect; LC is quiescent during REM sleep and is activated by sleep deprivation | $\alpha_2$ antagonists (idazoxan, mirtazapine) have been shown to exert AD and anxiolytic effects; use of an $\alpha_2$ antagonist during REM sleep is quite novel; human studies are just under way |

**Appendix 2:** Potential targets for the development of new treatments for mood and anxiety disorders *(continued)*

| Molecule | Hypothesized involvement in mood or anxiety disorders or Rx | Function plausibly relevant to mood or anxiety disorders | Findings from animal models | Direct or surrogate human evidence | Observations from clinical treatment studies |
|---|---|---|---|---|---|
| CRH antagonists | Enhanced throughput of CRH receptors may mediate some of the signs and symptoms of depression and anxiety; CRH antagonists may be effective ADs and/or anxiolytics | CRH regulates NE LC firing; CRH is increased by stress; CRH receptors are well placed to regulate many of the neurovegetative symptoms of depression | Agonists reproduce depression- and anxiety-like behaviors in rodents; the orally active CRH antagonist antalarmin significantly reduces fear and anxiety responses in nonhuman primates | HPA axis dysregulation has been found in depression; CSF and postmortem brain studies in depression are supportive | Positive effects seen in the initial study; however, the study was stopped because of likely mechanism-unrelated side effects; several other agents at various stages of development |

**Appendix 2:** Potential targets for the development of new treatments for mood and anxiety disorders *(continued)*

| **Molecule** | **Hypothesized involvement in mood or anxiety disorders or Rx** | **Function plausibly relevant to mood or anxiety disorders** | **Findings from animal models** | **Direct or surrogate human evidence** | **Observations from clinical treatment studies** |
|---|---|---|---|---|---|
| Short-term treatment with GR antagonists | Hypercortisolemia may play an important role in the pathophysiology and/or deleterious long-term consequences of mood and anxiety disorders | Hippocampal atrophy mediated in part by hypercortisolemia; other brain areas may be affected; diabetes, bone mineral density are also affected by hypercortisolemia and are found at an increased rate in patients with mood disorders | Injection of a GR antagonist into the dentate gyrus attenuates the acquisition of learned helpless behavior; transgenic and knockout mice exhibit some symptoms of anxiety and depression; ADs exert complex effects on GR expression and function | Abundant data exist demonstrating HPA axis activation in mood and anxiety disorders, especially in severely ill patients | Preliminary studies of mifepristone in psychotic depression are very encouraging; larger studies are under way |

**Appendix 2:** Potential targets for the development of new treatments for mood and anxiety disorders *(continued)*

| Molecule | Hypothesized involvement in mood or anxiety disorders or Rx | Function plausibly relevant to mood or anxiety disorders | Findings from animal models | Direct or surrogate human evidence | Observations from clinical treatment studies |
|---|---|---|---|---|---|
| NK-1 antagonists (substance P receptors) | Enhanced NK-1 function in depression (?); antagonists may be effective treatments for depression or anxiety | NK-1 plays an important role in mediating pain (? "psychic pain"); activation of NK-1 receptors reduces 5-HT neurotransmission | Efficacy in animal models of depression; stress regulates redistribution of NK-1 receptors; NK-1 knockout or NK-1 antagonism reduces anxiety-like behavior in certain models | No strong direct supportive evidence | Initial clinical studies yielded positive results in treating anxiety and depression; subsequent replications failed; awaiting more definitive studies |

**Appendix 2:** Potential targets for the development of new treatments for mood and anxiety disorders *(continued)*

| Molecule | Hypothesized involvement in mood or anxiety disorders or Rx | Function plausibly relevant to mood or anxiety disorders | Findings from animal models | Direct or surrogate human evidence | Observations from clinical treatment studies |
|---|---|---|---|---|---|
| CCK antagonists | Dysregulation of CCK could contribute to panic attacks, anorexia, or other manifestations of anxiety | CCK inhibits feeding and increases anxiety-like behavior | Data are mixed; differences in pharmacological agents, animal strain, and paradigms used may account for variable results | CCK agonists are anxiogenic and panicogenic in both patients with panic disorder and healthy volunteers | CCK antagonist CI-988 has not shown significant efficacy in clinical trials |
| NPY receptor agonists | NPY may serve as an endogenous anti-stress, anxiolytic agent; NPY agonists may be efficacious for certain symptoms of depression or anxiety | NPY may counter many of the deleterious effects of CRH and stresses | Efficacy has been shown in animal models of anxiety; NPY knockout shows reduced anxiety in certain models; ADs and lithium may increase NPY expression | CSF NPY may be low in depression; ECT increases CSF NPY-like immunoreactivity | No clinical studies to date |

**Appendix 2:** Potential targets for the development of new treatments for mood and anxiety disorders *(continued)*

| Molecule | Hypothesized involvement in mood or anxiety disorders or Rx | Function plausibly relevant to mood or anxiety disorders | Findings from animal models | Direct or surrogate human evidence | Observations from clinical treatment studies |
|---|---|---|---|---|---|
| $GABA_A$ subunit–selective BZD agonists | Certain subunits may be responsible for anxiolytic effects of BZDs, while other subunits may mediate side effects (sedation, amnesia, addiction) | GABAergic transmission provides inhibition to a variety of regions and cell types; the α2 subunit is highly expressed in the striatum, cortex, and hippocampus | Point mutation of the $\alpha_2$ BZD-binding site blocks diazepam anxiolysis in animal models; $\alpha_2/\alpha_3$–preferring agonists (L-838417, SL651498) have been shown to block anxiety-like behavior without sedation or ataxia | Possible subunit-specific expression differences in alcoholism | There are no reports of $\alpha_2$-targeting drugs yet; however, the sedative zaleplon, which preferentially activates $\alpha_1$-containing receptors, has fewer side effects than conventional BZDs, suggesting the usefulness of this strategy |

**Appendix 2:** Potential targets for the development of new treatments for mood and anxiety disorders *(continued)*

| Molecule | Hypothesized involvement in mood or anxiety disorders or Rx | Function plausibly relevant to mood or anxiety disorders | Findings from animal models | Direct or surrogate human evidence | Observations from clinical treatment studies |
|---|---|---|---|---|---|
| NMDA antagonists | Enhanced throughput of the NMDA receptor may contribute to brain regional volumetric changes observed in depression; NMDA antagonists may possess AD and anxiolytic efficacy | NMDA receptors are key regulators of many forms of synaptic plasticity; play an important role in stress-induced hippocampal atrophy and reduction of neurogenesis; implicated in many forms of cell atrophy and death | NMDA antagonists block stress-induced cell atrophy/ reduction of neurogenesis; many ADs regulate NMDA receptor subunit expression; NMDA antagonists are efficacious in certain animal models of depression and anxiety | Very indirect—regional volumetric reductions in mood disorders; evidence for glial and neuronal loss/atrophy in mood disorders | Amantadine and, especially, lamotrigine have AD efficacy; preliminary results suggest that ketamine may also have AD efficacy; studies with other NMDA antagonists being planned; side effects can be significant |

**Appendix 2:** Potential targets for the development of new treatments for mood and anxiety disorders *(continued)*

| Molecule | Hypothesized involvement in mood or anxiety disorders or Rx | Function plausibly relevant to mood or anxiety disorders | Findings from animal models | Direct or surrogate human evidence | Observations from clinical treatment studies |
|---|---|---|---|---|---|
| AMPA potentiators | AMPA receptors known to activate MAP kinase cascades and increase plasticity | AMPA receptors play important roles in neuronal functioning and plasticity | AMPA-potentiating agents have shown efficacy in animal models of depression; an ampakine (CX516) has been shown to produce a marked facilitation of performance in a memory task in rats | Very indirect—impairment of neuronal plasticity and cellular resilience | No studies as of yet in mood disorders; preliminary human studies suggest a positive memory encoding effect in certain spheres; benefical effects seen on measures of attention and memory when added to clozapine in schizophrenia |

**Appendix 2:** Potential targets for the development of new treatments for mood and anxiety disorders *(continued)*

| **Molecule** | **Hypothesized involvement in mood or anxiety disorders or Rx** | **Function plausibly relevant to mood or anxiety disorders** | **Findings from animal models** | **Direct or surrogate human evidence** | **Observations from clinical treatment studies** |
|---|---|---|---|---|---|
| Ligands for metabotropic glutamate receptors | Glutamate excitotoxicity has been implicated in stress-induced atrophy | Metabotropic glutamate receptors modulate glutamate signaling | mGluR5 antagonism shows anxiolytic and antinociceptive efficacy in animal studies; similar results have also been reported with group II mGluR agonists | None | No data have been reported to date |

**Appendix 2:** Potential targets for the development of new treatments for mood and anxiety disorders *(continued)*

| Molecule | Hypothesized involvement in mood or anxiety disorders or Rx | Function plausibly relevant to mood or anxiety disorders | Findings from animal models | Direct or surrogate human evidence | Observations from clinical treatment studies |
|---|---|---|---|---|---|
| PDE4 inhibitors | Reduced throughput of the cAMP signaling cascade may be involved in depression; enhancement of cAMP signaling may be AD | Enhances cAMP signaling and downstream gene expression, as well as synaptic plasticity and cell survival | PDE inhibitors effective in some models of depression; antidepressants enhance cAMP mediated signaling | Postmortem brain studies suggest a potential impairment of cAMP signaling cascade in depression (but not bipolar disorder) | Preliminary early clinical studies suggested AD efficacy of rolipram; newer clinical studies with PDE inhibitors as AD adjuncts are under way |
| MAP kinase phosphatase inhibitors | Enhancing neurotrophic factor signaling by inhibiting the turn-off reactions may be efficacious in the treatment of depression | MAP kinase signaling cascades are critical mediators of the effects of neurotrophic factors (e.g., BDNF) and play important roles in synaptic and structual plasticity | ADs and lithium increase BDNF expression; valproate activates the MAP kinase cascade | Very indirect—regional volumetric reductions in mood disorders; evidence for glial and neuronal loss/atrophy in mood disorders | No clinical studies with specific agents have been reported to date |

**Appendix 2:** Potential targets for the development of new treatments for mood and anxiety disorders *(continued)*

| Molecule | Hypothesized involvement in mood or anxiety disorders or Rx | Function plausibly relevant to mood or anxiety disorders | Findings from animal models | Direct or surrogate human evidence | Observations from clinical treatment studies |
|---|---|---|---|---|---|
| Isozyme-selective PKC inhibitors | Enhancement of PKC activity may play a role in the symptomatology of mania and/or anxiety; PKC inhibitors may be antimanic and anxiolytic | PKC isozymes play a major role in regulating neuronal firing and neurotransmitter release; may play important roles in psychostimulant-mediated catecholamine release | Lithium and valproate, on chronic administration, reduce the levels of PKC $\alpha$ and $\varepsilon$; certain biochemical and behavioral effects of psychostimulants are attenuated by PKC inhibitors, whereas BZD effects are potentiated | A postmortem brain study and human platelet studies suggest activation of PKC isozymes in bipolar disorder/mania; platelet studies also suggest normalization with lithium treatment | A preliminary study suggests that tamoxifen (an estrogen receptor antagonist and PKC inhibitor) has antimanic efficacy; larger clinical studies with tamoxifen are under way; valproate has shown efficacy in panic disorder and PTSD |

**Appendix 2:** Potential targets for the development of new treatments for mood and anxiety disorders *(continued)*

| Molecule | Hypothesized involvement in mood or anxiety disorders or Rx | Function plausibly relevant to mood or anxiety disorders | Findings from animal models | Direct or surrogate human evidence | Observations from clinical treatment studies |
|---|---|---|---|---|---|
| GSK-3 inhibitors, β-catenin upregulators | GSK-3 inhibitors and β-catenin upregulators may have mood stabilizing effects | GSK-3 plays an important role in structural plasticity and regulates cell death pathways | Lithium inhibits GSK-3 and upregulates β-catenin; VPA upregulates β-catenin likely via GSK-3 and non-GSK-3 mechanisms; lithium, VPA, and lamotrigine protect against GSK-3 overexpression–induced cell death | Very indirect—regional volumetric reductions in mood disorders; evidence for glial and neuronal loss/atrophy in mood disorders | Development of CNS-penetrant, selective small-molecule GSK-3 inhibitors is under way |

**Appendix 2:** Potential targets for the development of new treatments for mood and anxiety disorders *(continued)*

| Molecule | Hypothesized involvement in mood or anxiety disorders or Rx | Function plausibly relevant to mood or anxiety disorders | Findings from animal models | Direct or surrogate human evidence | Observations from clinical treatment studies |
|---|---|---|---|---|---|
| bcl-2 upregulators | Upregulating bcl-2 may exert trophic effects and enhance cellular resilience in the treatment of mood disorders | One of the major cell survival signals, and a major downstream effector of neurotrophic factors; likely plays an important role in neurite outgrowth, neurogenesis, and other forms of neuroplasticity | Lithium and VPA, on chronic administration, robustly upregulate bcl-2 levels and exert neuroprotective effects | Preliminary postmortem brain studies suggest the possible involvement of bcl-2 in mood disorders; lithium increases gray matter volumes in brain areas of reported atrophy in humans | Lithium and VPA are effective mood stabilizers; no selective CNS bcl-2 upregulators are currently available; however, pramipexole, in addition to having dopaminergic effects, upregulates bcl-2; positive AD effects in preliminary studies; larger studies are under way |

**Appendix 2:** Potential targets for the development of new treatments for mood and anxiety disorders *(continued)*

| Molecule | Hypothesized involvement in mood or anxiety disorders or Rx | Function plausibly relevant to mood or anxiety disorders | Findings from animal models | Direct or surrogate human evidence | Observations from clinical treatment studies |
|---|---|---|---|---|---|

*Note.* See text for references. AD = antidepressant; AMPA = (*R,S*)-α-3-hydroxyl-5-methyl-4-isoxazolepropionic acid; BDNF = brain-derived neurotrophic factor; BZD = benzodiazepine; CCK = cholecystokinin; CNS = central nervous system; CREB = cAMP response element binding protein; CRH = corticotropin-releasing hormone; CSF = cerebrospinal fluid; ECT = electroconvulsive therapy; GABA = γ-aminobutyric acid; $GABA_A$ = γ-aminobutyric $acid_A$ receptor; GR = glucocorticoid receptor; GSK-3 = glycogen synthase kinase–3; HPA = hypothalamic-pituitary-adrenal axis; 5-HT = serotonin; $5\text{-}HT_{1A}$ = $serotonin_{1A}$ receptor; $5\text{-}HT_{1B}$ = $serotonin_{1B}$ receptor; $5\text{-}HT_2$ = $serotonin_2$ receptor; $5\text{-}HT_3$ = $serotonin_3$ receptor; $5\text{-}HT_7$ = $serotonin_7$ receptor; LC = locus coeruleus; MAP = mitogen activated protein; mGluR5 = metabotropic glutamate 5 receptor; NE = norepinephrine; NK-1 = neurokinin-1 receptor; NMDA = *N*-methyl-D-aspartate; NPY = neuropeptide Y; PDE = phosphodiesterase; PDE4 = phosphodiesterase 4 isoforms; PET = positron emission tomography; PKC = protein kinase C; PTSD = posttraumatic stress disorder; REM = rapid eye movement; VPA = valproate.

# Index

*Page numbers in* **boldface** *type refer to tables or figures.*